Game Theory Solutions for the Internet of Things:

Emerging Research and Opportunities

Sungwook Kim
Sogang University, South Korea

A volume in the Advances in Web Technologies and Engineering (AWTE) Book Series

www.igi-global.com

Published in the United States of America by
IGI Global
Information Science Reference (an imprint of IGI Global)
701 E. Chocolate Avenue
Hershey PA 17033
Tel: 717-533-8845
Fax: 717-533-8661
E-mail: cust@igi-global.com
Web site: http://www.igi-global.com

Library of Congress Cataloging-in-Publication Data

Names: Kim, Sungwook, author.
Title: Game theory solutions for the internet of things : emerging research and opportunities / by Sungwook Kim.
Description: Hershey PA : Information Science Reference, [2017] | Includes bibliographical references.
Identifiers: LCCN 2016048911| ISBN 9781522519522 (hardcover) | ISBN 9781522519539 (ebook)
Subjects: LCSH: Internet of things. | Game theory.
Classification: LCC TK5105.8857 .K56 2017 | DDC 004.67/80151932--dc23 LC record available at https://lccn.loc.gov/2016048911

This book is published in the IGI Global book series Advances in Web Technologies and Engineering (AWTE) (ISSN: 2328-2762; eISSN: 2328-2754)

British Cataloguing in Publication Data
A Cataloguing in Publication record for this book is available from the British Library.

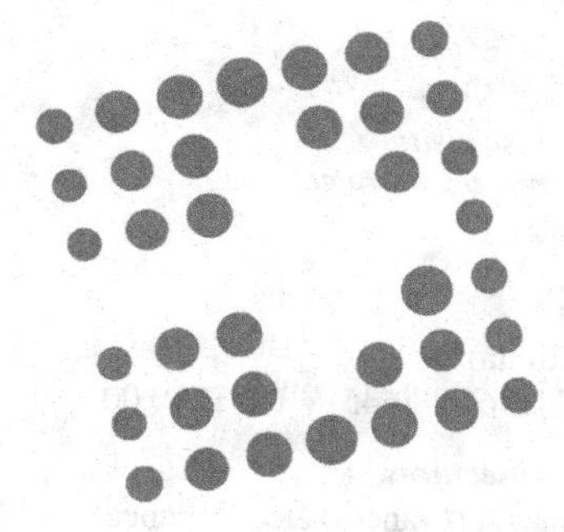

Advances in Web Technologies and Engineering (AWTE) Book Series

ISSN:2328-2762
EISSN:2328-2754

Editor-in-Chief: Ghazi I. Alkhatib, The Hashemite University, Jordan
and David C. Rine, George Mason University, USA

Mission

The **Advances in Web Technologies and Engineering (AWTE) Book Series** aims to provide a platform for research in the area of Information Technology (IT) concepts, tools, methodologies, and ethnography, in the contexts of global communication systems and Web engineered applications. Organizations are continuously overwhelmed by a variety of new information technologies, many are Web based. These new technologies are capitalizing on the widespread use of network and communication technologies for seamless integration of various issues in information and knowledge sharing within and among organizations. This emphasis on integrated approaches is unique to this book series and dictates cross platform and multidisciplinary strategy to research and practice.

The **Advances in Web Technologies and Engineering (AWTE) Book Series** seeks to create a stage where comprehensive publications are distributed for the objective of bettering and expanding the field of web systems, knowledge capture, and communication technologies. The series will provide researchers and practitioners with solutions for improving how technology is utilized for the purpose of a growing awareness of the importance of web applications and engineering.

Coverage

- Web user interfaces design, development, and usability engineering studies
- Integrated Heterogeneous and Homogeneous Workflows and Databases within and Across Organizations and with Suppliers and Customers
- Mobile, location-aware, and ubiquitous computing
- Competitive/intelligent information systems
- Knowledge structure, classification, and search algorithms or engines
- IT readiness and technology transfer studies
- Security, integrity, privacy, and policy issues
- Data analytics for business and government organizations
- Software agent-based applications
- Data and knowledge validation and verification

IGI Global is currently accepting manuscripts for publication within this series. To submit a proposal for a volume in this series, please contact our Acquisition Editors at Acquisitions@igi-global.com or visit: http://www.igi-global.com/publish/.

The Advances in Web Technologies and Engineering (AWTE) Book Series (ISSN 2328-2762) is published by IGI Global, 701 E. Chocolate Avenue, Hershey, PA 17033-1240, USA, www.igi-global.com. This series is composed of titles available for purchase individually; each title is edited to be contextually exclusive from any other title within the series. For pricing and ordering information please visit http://www.igi-global.com/book-series/advances-web-technologies-engineering/37158. Postmaster: Send all address changes to above address.

Titles in this Series

For a list of additional titles in this series, please visit:
http://www.igi-global.com/book-series/advances-web-technologies-engineering/37158

Design Solutions for Improving Website Quality and Effectiveness
G. Sreedhar (Rashtriya Sanskrit Vidyapeetha (Deemed Universit), India)
Information Science Reference • ©2016 • 423pp • H/C (ISBN: 9781466697645) • US $220.00

Handbook of Research on Redesigning the Future of Internet Architectures
Mohamed Boucadair (France Télécom, France) and Christian Jacquenet (France Téléco, France)
Information Science Reference • ©2015 • 621pp • H/C (ISBN: 9781466683716) • US $345.00

Artificial Intelligence Technologies and the Evolution of Web 3.0
Tomayess Issa (Curtin University, Australia) and Pedro Isaías (Universidade Aberta (Portuguese Open University), Portugal)
Information Science Reference • ©2015 • 422pp • H/C (ISBN: 9781466681477) • US $225.00

Frameworks, Methodologies, and Tools for Developing Rich Internet Applications
Giner Alor-Hernández (Instituto Tecnológico de Orizaba, Mexico) Viviana Yarel Rosales-Morales (Instituto Tecnológico de Orizaba, Mexico) and Luis Omar Colombo-Mendoza (Instituto Tecnológico de Orizaba, Mexico)
Information Science Reference • ©2015 • 349pp • H/C (ISBN: 9781466664371) • US $195.00

Handbook of Research on Demand-Driven Web Services Theory, Technologies, and Applications
Zhaohao Sun (University of Ballarat, Australia & Hebei Normal University, China) and John Yearwood (Federation University, Australia)
Information Science Reference • ©2014 • 474pp • H/C (ISBN: 9781466658844) • US $325.00

Evaluating Websites and Web Services Interdisciplinary Perspectives on User Satisfaction
Denis Yannacopoulos (Technological Educational Institute of Piraeus, Greece) Panagiotis Manolitzas (Technical University of Crete, Greece) Nikolaos Matsatsinis (Technical University of Crete, Greece) and Evangelos Grigoroudis (Technical University of Crete, Greece)
Information Science Reference • ©2014 • 354pp • H/C (ISBN: 9781466651296) • US $215.00

Solutions for Sustaining Scalability in Internet Growth
Mohamed Boucadair (France Telecom-Orange Labs, France) and David Binet (France Telecom, France)
Information Science Reference • ©2014 • 288pp • H/C (ISBN: 9781466643055) • US $190.00

For an enitre list of titles in this series, please visit:
http://www.igi-global.com/book-series/advances-web-technologies-engineering/37158

www.igi-global.com

701 East Chocolate Avenue, Hershey, PA 17033, USA
Tel: 717-533-8845 x100 • Fax: 717-533-8661
E-Mail: cust@igi-global.com • www.igi-global.com

Table of Contents

Epigraph vi

Preface vii

Acknowledgment xiii

Chapter 1
Basic Concepts of Internet of Things and Game Theory 1

Chapter 2
Cloud-Based IoT System Control Problems 13

Chapter 3
IoT System Resource Sharing Mechanisms 78

Chapter 4
New Game Paradigm for IoT Systems 101

Chapter 5
Energy-Aware Network Control Approaches 148

Chapter 6
Developing IoT Applications for Future Networks 171

Related Readings 202

About the Author 219

Index 220

Epigraph

The aim of research is the flowing of persons. Sometimes, you seem concerned mostly with the perfectioning of things.
In the land of research, you cannot walk by the light of someone else's lamp. You want to borrow mine. I would rather teach you how to make your own. Teaching only takes place when learning does. Learning only takes place when you teach something to yourself.
I hope, you learn that your mentor isn't someone you can lean on but someone who rids you of your tendency to lean. The day you follow someone you cease to follow Truth.
There was once a student who never became a mathematician because he blindly believed the answers he found at the back of his math textbook and, ironically, the answers were correct.

- Anthony de Mello (1985)

***

Ad majorem Dei gloriam et
Ora pro nobis peccatoribus, nunc et in hora mortis nostrae

Preface

One of the most fascinating trends today is the emergence of low-cost microcontrollers that are sufficiently powerful to connect to the Internet. They are the key to the Internet of Things (IoT), where all kinds of devices become the Internet's interface to the physical world. Usually, IoT is a globally interconnected collection of devices, systems and services that are being coordinated either manually or automatically to operate and orchestrate useful functions for the improvement of quality of life. In particular, IoT refers to an emerging paradigm that seamlessly integrates a large number of smart objects with the Internet interlinking the physical and the cyber worlds and keeping them in a tight and continuous interaction. In this envisioned paradigm, billions of smart objects will be immersed in the environment, sensing, interacting, and cooperating with each other to enable efficient services that will bring tangible benefits to the environment, the economy and the society as a whole. In addition, they will be extremely diverse and heterogeneous in terms of resource capabilities, lifespan and communication technologies, further complicating the scenario.

IoT is a new revolution of the Internet that is rapidly gathering momentum driven by the advancements in sensor networks, mobile devices, wireless communications, networking and cloud technologies. Experts forecast that by the year 2050 there will be a total of 50 billion things connected to the internet. According to research reports, currently more than 90 percent of things are not yet connected to the Internet. This provides massive opportunity for global connectivity by utilizing Internet next generation IPv6 addressing. Globally connected things can help enable automation, machine-to-machine communication and data analysis at an unprecedented scale. The pervasive nature of the web services accessible through readily available browsers and web clients provides a simple interface that can be used to semi-autonomously control and interact with the connected things. The apparent benefit of IoT is a collective intelligence that is beyond the capability of each non-connected thing. The IoT will also be riding on the capability of cloud computing and big data for relevant information processing, analysis and intelligence. The benefits to society of having properly functional IoT is immense. Scientists, engineers, technologists, sociologists, businessmen, policy makers, economists, artists and society as a whole should play their parts, thus the IoT can help improve our quality of life.

To effectively handle the IoT system, new problems and challenges arise spanning different areas: architecture, communication, addressing, routing, data and network management, resource allocation, quality of services, power and energy storage, security and privacy, etc. To realize the full potential of the IoT paradigm, it is necessary to address several challenges and develop suitable conceptual and technological solutions for tackling them. These include development of scalable architecture, moving from closed systems to open systems, dealing with privacy and ethical issues involved in data sensing; storage, processing, and actions; designing interaction protocols; autonomic management; communication protocol; smart objects and service discovery; programming framework; resource management; data and network management; power and energy management; and governance.

The complexity of IoT challenges many research domains, including decision models, optimization, data analysis, intelligent system, consumer behaviors, e-commerce, design science, cloud computing, semantics web, etc. The goal of these issues is to foster theoretical development and innovative practices on new theories, methodologies, technologies, and applications related to IoT. Classical network management approaches are not sufficient to solve these unprecedented issues, and need to be revised to address the complex requirements imposed by IoT. This paves the way for the development of intelligent algorithms, novel network paradigms and new services. Recently, game theory has become a useful tool for modeling and studying various IoT systems. Game theory is a mathematical framework to analyze complex interactions of cooperative or competing decision makers taking into account their preferences and requirements. Despite these early efforts and many other contributors in the history of game theory, it is widely accepted that the origin of the formal study of game theory began with John von Neumann and Oskar Morgenstern's book, *Theory of Games and Economic Behavior* (published in 1944). This pioneering work focused on finding unique strategies that allowed players to minimize their maximum losses by considering, for every possible strategy of their own, all the possible responses of other players. During the 1940s, game theory emerged from the fields of mathematics and economics to provide a revolutionary new method of analysis. Therefore, originally, game theory was invented to explain complicated economic behavior. The golden age of game theory occurred in the 1950s and 1960s when researchers focused on finding sets of strategies, known as equilibria, to solve a game if all players behaved rationally. The most famous of these is the Nash equilibrium proposed by John Nash. Nash also made significant contributions to bargaining theory and examined cooperative games where threats and promises are fully binding and enforceable. Today game theory provides a language for discussing conflict and cooperation. With its effectiveness in studying complex dynamics among players, game theory has been widely applied in economics, political science (e.g., voter behavior), sociology (network games), law (antitrust), computer science (defending networks against attacks), military science, biology (evolutionary strategies), pure philosophy, with occasional appearances in psychology, religion (recall Aumann's Talmud paper), physics (quantum games), telecommunications, and logistics, etc. In these fields, game theory has been employed to achieve socially optimal equilibrium.

In nowadays, game theory has been widely recognized as an important tool in the research area of IoT. The present research has witnessed a huge explosion of interest in issues that intersect IoT and game theory. A promising potential application of game theory in IoT is the area of computation offloading, cloud computing, cognitive radio, resource sharing, power control, routing, load balancing, multi-commodity flow, capacity allocation, incentive mechanism for cooperation between nodes and quality of service provisioning. In addition, new problems are constantly propounded in various areas such as network security and green energy harvesting. In recent years, these issues have become an increasingly important part of IoT applications. This book systematically introduces and explains the application of game theory in IoT management. It provides a comprehensive technical guide covering introductory concepts, fundamental techniques, recent advances, and open issues in related topics. Therefore, an important component of this book is examples of algorithms, which will give the readers the opportunity to clearly understand to get game theoretic solutions for specific IoT system management. In particular, I have tried to write a comprehensive book that transfers knowledge through various game approaches, where the reader is provided the necessary guidance and knowledge to develop theoretical models for real-world IoT applications. This focus is on core concept that has impact on improving system efficiency. Concurrent development of practical applications that accompanies game theory within the book further enhances the learning process, in my opinion.

The book is comprising of a total of 6 chapters. Chapter 1 covers the basic concepts for IoT and game theory. In this chapter, game theory and various

IoT applications are briefly presented. In particular, this chapter provides an overview of IoT, building blocks of IoT, IoT enabling technologies, characteristics of IoT systems and IoT levels.

Chapter 2 introduces the concept of cloud based IoT systems. Cloud computing is an emerging research area that targets on providing services and satisfying customers' needs. This technology is probably the most promising technology to support IoT applications while simultaneously and successfully addressing all the aforementioned challenges and issues. It adds a new dimension to IoT model for meeting the customers' needs such as fast connection, easy for management, infrastructure reuse, and off load core network traffic. Cloud computing is expected to support a wide range of IoT applications, including device-to-device data sharing, wearable cognitive assistance, video editing and sharing, vehicular systems, and etc. In the current cloud-based IoT model, smart devices, such as sensors, smartphones, exchange information through the Internet to cooperate and provide services to users, which could be citizens, smart home systems, and industrial applications. Even though the cloud based IoT model describes a uniform, concise, and scalable solution for supporting IoT applications, the deployments of IoT applications on cloud, however, are facing the challenges originated from economic considerations, social concerns, technical limitations, and administrative issues. To address the arising new challenges and opportunities, I select five cloud based IoT control schemes to help both industry and academia research communities better understand the recent advances and potential research directions on the converging paths of cloud-based IoT and game theory. These schemes are covered, describing how to develop adaptable game models for various cloud services.

Chapter 3 is dedicated to the IoT System Resource Sharing Mechanisms. In this chapter, the schemes for sharing the bandwidth and computation capacity are introduced. The IoT platforms are anticipated to host trillions of interconnected objects in the coming decades. With the help of sensors, receivers, and big data analytics, the IoT is set to reinvent the world, giving us new ways to monitor, use, manage, automate, and play with objects. Therefore, the issue of IoT resource sharing is a hot research topic. This chapter describes the characteristics of IoT resource sharing mechanisms including various environments.

Chapter 4 presents new game paradigms for IoT Systems. They outline the organizational, structural and control issues that are commonly encountered in the IoT environments. In addition, they provide a detailed overview of the new challenges related to the new game models. The emergence of large-scale and decentralized heterogeneous IoT systems operating under dynamic and uncertain environments imposes new challenges in the design, analysis, and optimization of IoT systems. Among the different mathematical theories that can be used to study and develop such networks, game theory is very promising as it can capture many of the intrinsic features of these IoT systems. Therefore, this chapter aims to gather cutting-edge contributions that address and show the latest results on developing game-theoretic models for emerging IoT applications. I solicit high-quality original IoT control schemes on advanced game-theoretic paradigms that go beyond classical models and solution concepts, thereby making the vision of large-scale, heterogeneous, dynamic and self-adapting IoT systems a reality.

Chapter 5 focuses on the energy oriented IoT control approach. It has always been an integral part of IoT applications. Usually, IoT assigns unique identifications to individual objects, be they human beings, home appliances, animals, cars, phones, watches, animals, or objects we have yet to imagine. These uniquely identified objects are connected through the Internet in wired or wireless fashion. To effectively handle the interconnected objects, energy control is a hot research topic. Therefore, the aim of this chapter is to provide the discussion and technical presentations on the recent advances in energy-aware IoT control approaches, theory, application and implementation of protocols, algorithms, and services.

Chapter 6 discusses the developing IoT applications for future networks. Many applications with high social and business impact fall under the IoT umbrella, including personal healthcare, smart grid, surveillance, home automation, intelligent transportation, while it is expected that new ones will emerge once the enabling technologies reach a stable state. At the moment, one of the most important challenges is to design control protocols and algorithms for an efficient interconnection of smart objects, both between themselves and with the Internet. The other issue is to create value-added services in cross-domain applications, especially open and interoperable, enabled by the interconnection of things, machines, smart objects, in such a way that they can be integrated with current and new business and development processes. This chapter explains new and novel developing applications for future IoT systems.

This book can enable readers to grasp the fundamental concepts of game theory and IoT applications. Through the use of some state-of-the-art examples for various IoT control schemes, I will illustrate the wide varieties of topics and their potential future design directions. This point is equally relevant and applicable in today's world of IoT technology and game theory. It is the hope that this brief guide can shed light on the opportunities opened up by IoT, and is expected that diligent readers of this book can use these exercises to develop their own IoT control models. This book is also intended primarily for use in a network engineer, architect, or technician who has a working knowledge of IoT protocols and technologies. Especially, my focus is on getting the reader firmly on track to developing novel IoT control schemes. In addition, this book will provide practical advice on applying the knowledge to IoT applications for graduate students in computer science, mathematics and electrical engineering. Finally, this book is equally well suited for self-study technical professionals since it discusses engineering issues in algorithm design, as well as mathematical aspects. I expect the book to serve as a reference for systems architects, practitioners, researchers, and graduate-level students.

Acknowledgment

Writing a book is a both painful and enjoyable experience. It's just like climbing a high peak, step by step, accompanied with bitterness, hardships, frustration, encouragement and trust and with so many people's kind help. When I found myself at the top enjoying the beautiful scenery, I realized that it was, in fact, *human beings* that got me there. I can say that I've been surrounded by wonderful people. It will not be enough to express my gratitude in words to all those people who helped me. Therefore, I would like to give my deepest and most sincere thanks to God, who kindly sent these wonderful people to me.

I would still like to give my many, many thanks to all these people. First of all, I'd like to give my sincere thanks to my honorific academic supervisor, Prof. Pramod K. Varshney, who accepted me as his Ph.D. student at Syracuse University. Thereafter, he offered me so much advice, patiently supervising me, and always guiding me in the right direction. I've learned a lot from him, without his help I could not find the path of science, and knowledge.

My time at Sogang University was made enjoyable in large part due to the many friends and colleagues that became a part of my life. I am grateful for time spent with them. I also wish to thank my loving, supportive, encouraging, and patient graduate students. They are the best. I am so incredibly grateful for each and every one of my students. I am forever grateful to have them in my life. I know they are there and I appreciate that more than they could ever understand. Specially, I am grateful for Dr. Youngjae Park, who is my first Ph.D student, and my best friend. His support without any complaint or regret has enabled me to complete this book. I take this opportunity to express the profound gratitude from my deep heart to him.

I'd like to convey my heartfelt thanks to my parents, Sangjoon Kim and Jungsun Shin, and my elder brother, Jungwook Kim, for their faith in me and allowing me to be as ambitious as I wanted. It was under their watchful eye that I gained so much drive and an ability to tackle challenges head on.

They have a wonderful way of making each day brighter, and any moment shared with them makes the heart a little lighter. They are always around when needed to show how much they care. And that is why it means so much just knowing that they are there. I am deeply thankful to them for their love, support, and sacrifices. Without them, this book would never have been written. I dedicate this book to them. And, my grandparents, I hope they are looking down from somewhere with a smile. Words cannot truly express my deepest gratitude and appreciation to them. Definitely, they are remembered and accepted by me, now and forever.

Finally, and most importantly, praises and thanks to the God, the Almighty, for His showers of blessings throughout my research work to complete this book successfully. In the process of putting this book together I realized how true this gift of writing is for me. He given me the power to believe in my passion and pursue my dreams. I could never have done this without the faith I have in Him, the Almighty.

Sungwook Kim
Sogang University, South Korea

Chapter 1
Basic Concepts of Internet of Things and Game Theory

ABSTRACT

With the evolution of the Internet and related technologies, there has been an evolution of new paradigm, which is the Internet of Things (IoT). IoT is the network of physical objects, such as devices, embedded with electronics, software, sensors, and network connectivity that enables these objects to collect and exchange data. In the IoT, a large number of objects are connected to one another for information sharing, irrespective of their locations (Corcoran, 2016). Even though the IoT was defined at 1999, the concept of IoT has been in development for decades. As the technology and implementation of the IoT ideas move forward, different views for the concept of the IoT have appeared (Ma, 2011). Based on different views, in this book, the IoT is defined as a kind of modern technology, implicating machine to machine communications and person to computer communications will be extended to everything from everyday household objects to sensors monitoring the movement. Currently, we can see a few key areas of focus for the Internet of Things (IoT) that will require special attention over the course of the next decade on the part of computer science, energy technology, networks, wireless communication, and system platform. There are already a number of implementation case studies emerging from companies across a range of industry sectors.

DOI: 10.4018/978-1-5225-1952-2.ch001

CLOUD COMPUTING WITH IoT

In 2010s, it seems that we are seeing a new phenomenon; the data are leaving the computer and moving to the network. Using the cloud computing, we can realize that it is quite empowering to have all of our data available in one place and from any device. Once we start using these services, more of our day-to-day data tend to get sucked up into cyberspace. From the practical viewpoint, cloud computing and IoT are two very different technologies that are both already part of our life. Their adoption and use are expected to be more and more pervasive, making them important components of the future Internet. A novel paradigm where cloud and IoT are merged together is foreseen as disruptive and as an enabler of a large number of application scenarios. We call this new paradigm as CloudIoT, which is expected to disrupt both current and future Internet (Botta, 2016; Corcoran, 2016). The new applications, arising from this CloudIoT paradigm, open up new exciting directions for business and research; smart cities will enable more efficient public services and promote new business opportunities, ubiquitous healthcare applications will improve the quality of life for many patients (Botta, 2016). In this book, designing cloud oriented IoT mechanisms and the deployment of CloudIoT are considered as one of the main issues.

GREEN IoT

IoT devices will produce a lot of electronic waste and will also consume a significant amount of energy in order to execute different tasks. Besides energy consumption is acute in different heterogeneous IoT devices as it actively relates to cost and availability of the IoT network. However, energy is considered as valuable resource for IoT network, because the devices used for IoT applications are battery operated low power machines. Therefore, utilizing the energy in efficient way is the main goal of IoT network. This will eventually pose a challenge in near future to reduce the energy consumption and will also demand for new ways of developing a green communication across the IoT systems. Nowadays, new challenges for the energy-efficient IoT system design have been addressed. However, despite these efforts to improve the energy efficiency in IoT systems, it is not always practical or feasible under limited ambient energy availability and stringent form-factor constraints. Therefore, additional system-level techniques need to be developed for the IoT systems. For the future IoT, energy consumption has become a core issue and different algorithmic approaches have been initiated for different effective solutions like complementing hardware or different system-based approaches (Abedin, 2015).

QUALITY OF SERVICE IN IoT

The huge number of different links and interactions between IoT objects makes it a scalable complex system. In addition, some services in service-oriented IoT are required to be reconfigurable and composable for Quality of Service (QoS) aware services. Therefore, brings difficulties for satisfying the dynamic QoS requirements of services (Li, 2014). In this perspective, it is necessary to define service models, which can categorize IoT applications and then determine which QoS factors are necessary to satisfy the requirements of those services (Nef, 2012). Nowadays, a number of QoS models have been developed for traditional networks. However, the QoS management in IoT is still poorly studied. The definition of QoS in IoT is still not clear because the definition of service in IoT is not exactly the same, in which a service can be defined as the simple acquisition and processing of information and the decision making process in identification, communication, and so on. The traditional QoS attributes in terms of bandwidth, delay, jitter, and packet loss ratio are evidently inappropriate in IoT. In IoT, more QoS attributes are concerned, such as information accuracy, the network resources needed, required energy consumption, and the coverage of IoT. To solve the difficulties mentioned above, a new QoS model for service-oriented IoT is necessary (Nef, 2012).

RESOURCE MANAGEMENT FOR IoT SYSTEMS

In the IoT, multiple devices are active participants in business, information and social processes where they are enabled to interact and communicate among themselves and with the environment by exchanging data and information sensed about the environment, while reacting autonomously to the real world events (Nef, 2012). Therefore, the IoT contains a large number of different devices and heterogeneous networks, which make it difficult to satisfy different resource requirements and achieve rapid resource deployment. When entirely different and unexpected IoT devices would be asking for resources, resource allocation would be a challenge. Because it would be very difficult to decide how much a particular resource may be required by an entity or a particular IoT device. Depending upon the type, amount, and priority of service, resource allocation has to be mapped (Aazam, 2014). As a result, a large variety of resource management protocol are needed in order to improve IoT performance while gaining benefit from the deployed devices. This raises the question how the resources provided by the devices can be efficiently managed and provisioned. Related concepts like on-demand provisioning, elasticity, or resource pooling and sharing are already known from the IoT domain. The dominating performance factors in resource allocation are the IoT system's profit, user's utility, resource utilization, and QoS ensuring (Aazam, 2014; Nef, 2012).

FOG COMPUTING PARADIGM WITH IoT

In the past decade, the evolution toward 5G is featured by the explosive growth of traffic in the wireless network, due to the exponentially increased number of user devices. Compared to the 4G communication system, the 5G system should bring billions of user devices into wireless networks to demand high bandwidth connections. Therefore, system capacity and energy efficiency should be improved to get the great success of 5G communications (Hung, 2015; Park, 2016; Tandon, 2016). Fog computing is a promising solution to the mission critical tasks involving quick decision making and fast response. It is a distributed paradigm that provides cloud-like services to the network edge nodes. Instead of using the remoted cloud center, the fog computing technique leverages computing resources at the edge of networks based on the decentralized transmission strategies. Therefore, it can help overcome the resource contention and increasing latency. Due to the effective coordination of geographically distributed edge nodes, the fog computing approach can meet the 5G application constraints, i.e., location awareness, low latency, and supports for mobility or geographical distribution of services. The most frequently referred use cases for the fog computing concept are related to the IoT (Borylo, 2016; Dastjerdi, 2016).

UNDERWATER IoT

The commercial benefits of the IoT are clear for a wide range of industries. However, some of the shiny promise of interconnectedness can get washed away at the shoreline, given the technical challenges of monitoring and communicating underwater. On the earth, over 70% of the area covered by water in the form of rivers, canals, seas, and oceans. Therefore, the exploration of underwater acoustic transmission has recently attracted much attention due to its significant ability in distributed tactical surveillance, disaster preventing, mine reconnaissance and environment monitoring. Recent advances in communication technologies have led the possibilities to develop Underwater Sensor Networks (USNs). Based on the enabling technology for underwater explorations, USN is a network, consisting of sink nodes and autonomous micromechanical sensors. Sensors are spatially distributed in the water to sense the water-related information, and connected wirelessly through acoustic signal in the underwater environment. The sink on the surface collects the sensed data from the underwater sensors and transmits this collected information to the monitoring center via satellite for further analysis. Even though USN has emerged as a promising networking technique for various underwater applications, the USN operation presents its own challenges in terms of dynamic structure, rapid energy consumption, narrow bandwidth and long propagation delay (Ghoreyshi, 2016; Shah, 2016; Wahid, 2012).

RADIO FREQUENCY ENERGY HARVESTING FOR IoT SYSTEMS

Wireless communication network is becoming more and more important, and has recently attracted a lot of research interest. Compared with wireline communication, wireless communication has lower cost, which is easier to be deployed. With the development of Internet of Things (IoT) and embedded technology, wireless communication will be applied in more comprehensive scopes. Sometimes, wireless devices perform complex task with portable batteries. However, batteries present several disadvantages like the need to replace and recharge periodically. As the number of electronic devices continues to increase, the continual reliance on batteries can be both cumbersome and costly (Lim, 2013; Park, 2012). Recently, Radio Frequency (RF) energy harvesting has been a fast growing topic. The RF energy harvesting is developed as the wireless energy transmission technique for harvesting and recycling the ambient RF energy that is widely broadcasted by many wireless systems such as mobile communication systems, Wi-Fi base stations, wireless routers, wireless sensor networks and wireless portable devices (Khansalee, 2014). Therefore, this technique becomes a promising solution to power energy-constrained wireless networks while allowing the wireless devices to harvest energy from RF signals. In RF energy harvesting, radio signals with frequency range from 300 GHz to as low as 3 kHz are used as a medium to carry energy in a form of electromagnetic radiation. With the increasingly demand of RF energy harvesting/charging, commercialized products, such as Powercaster and Cota system, have been introduced in the market (Lu, 2015).

SOCIAL NETWORK SERVICE WITH IoT

With the advance of the IoT, Social Network Service (SNS) has attracted billions of Internet users from all over the world in the past few years. SNS connects people to provide online communication and collaboration environment beyond the geographic limitations. It is considered to be a representative of the new generation Internet applications, and many specialized SNSs have emerged (Du, 2012). Usually, the main goal of SNS is to seek reciprocal value creation to increase the productivity, quality, and opportunities of online services. To satisfy this goal, SNS users will expect more application services to fulfill their needs beyond fundamental service functions. Therefore, how to link the needs of users and shape the designs for better service utilization is an important issue in SNS research fields (Hwang, 2012). For the interoperability of the SNS and cloud services, a new concept, Social Cloud (SC) was introduced based on the notion of resource and service collaboration. SC is a novel scalable computing model where resources are beneficially shared among a group of Social Network (SN) users. From (Chard, 2012), the formal definition of SC is like as; *A social cloud is a resource and service sharing framework utilizing relationships established between members of a social network.* It is a resource and service sharing framework utilizing relationships established between SNS users and CC provides. Under a dynamic IoT environment, the idea of SC has been gaining importance because of their potential for the system efficiency (Kim, 2016).

MOBILE CROWD SENSING WITH IoT

With the development of IoT and embedded technology, the remote intelligent monitoring system will be applied in more comprehensive scopes. Therefore, ubiquity of internet-connected portable devices is enabling a new class of applications to perform sensing tasks in the real world. Among mobile devices, smartphones have evolved as key electronic devices for communications, computing, and entertainment, and have become an important part of people's daily lives. Most of current mobile phones are equipped with a rich set of embedded sensors, which can also be connected to a mobile phone via its Bluetooth interface. These sensors can enable attractive sensing applications in various domains such as environmental monitoring, social network, healthcare, transportation, and safety (Sheng, 2014). Mobile Crowd Sensing (MCS) refers to the technology that uses mobile devices, i.e., smartphones, to collect and analyze the information of people and surrounding environments (An, 2015). Based on this information, we can analyze statistical characteristics of group behaviors, reveal hidden information of social activity patterns, and finally provides useful information and services to end users. By involving anyone in the process of sensing, MCS greatly extends the service of IoT and builds a new generation of intelligent networks that interconnect things-things, things-people and people-people. Therefore, to provide a new way of perceiving the world, MCS has a wide range of potential applications (An, 2015).

GAME THEORY: NEW SOLUTION PARADIGM FOR IoT

The IoT environment, i.e., multi-user, multi-technology, multi-application, and multi-provider environment, requires a development of new solution concept. Usually, individual network agents locally make control decisions to maximize their profits in a distributed manner. This situation leads us into game theory. Game theory is a field of applied mathematics that provides an effective tool in modeling the interaction among independent decision makers (Gibbons, 1992a, 1992b; Osborne, 1994). It can describe the possibility to react to the actions of the other decision makers and analyze the situations of conflict and cooperation in real world. The rational decision makers, referred to as 'players' in a game model, try to maximize their expected benefits through strategy set. Many applications of game theory are related to economics, but it is also a powerful tool to model a wider range of real life situations, such as political science, sociology, psychology, biology, computer science and telecommunications so on, where conflict and cooperation exist. Recently, multi-criteria decision making algorithms in IoT have been added to this list (Bruin, 2005; Han, 2011; Liang, 2013; Shubik, 1973; Wooldridge, 2012).

Generally, game theory classified into two branches: non-cooperative and cooperative game theory (Nash, 1950, 1953). Non-cooperative game theory studies the strategic choices resulting from the interactions among competing players, where each player improving its strategy independently for improving its own performance or reducing its loss. Nash equilibrium (NE) is a one of the solution for non-cooperative game (Roy, 2010). Cooperative game theory provides analytical tools to study the behavior of rational players when they cooperate make a coalition to strengthen their positions in the game. Bargaining is a well-known solution for cooperative games (Kim, 2014, MacKenzie, 2001; Rubinstein, 1982). In this book, we suggest different game models used to study the interaction between the competitive and/or cooperative behavior that can be identified among network agents in IoT environments. The proposed game models have been successfully applied to typical IoT management problems, like cloud computation, data offloading, spectrum resource allocation, power control and QoS provisioning problems.

In summary, the objective of this book is to provide a didactic approach to studying game theory for the hot research issues in the IoT system management. Through the use of some state-of the-art game models for different IoT control issues, we will provide a comprehensive technical guide and potential future design directions.

REFERENCES

Aazam, M., Khan, I., Alsaffar, A. A., & Huh, E. N. (2014). Cloud of Things: Integrating Internet of Things and cloud computing and the issues involved. *IBCAST*, *8*(12), 414–419.

Abedin, S. F., Alam, M. G. R., Haw, R., & Hong, C. S. (2015). A system model for energy efficient green-IoT network. *IEEE ICOIN*, *2015*, 177–182.

An, J., Gui, X., Yang, J., Sun, Y., & He, X. (2015). Mobile Crowd Sensing for Internet of Things: A Credible Crowdsourcing Model in Mobile-Sense Service. *IEEE BigMM*, *2015*, 92–99.

Borylo, P., Lason, A., Rzasa, J., Szymanski, A., & Jajszczyk, A. (2016). Energy-aware fog and cloud interplay supported by wide area software defined networking. *IEEE ICC*, *2016*, 1–7.

Botta, A., Donato, W., Persico, V., & Pescapé, A. (2016). Integration of Cloud computing and Internet of Things: A survey. *Future Generation Computer Systems*, *56*, 684–700. doi:10.1016/j.future.2015.09.021

Bruin, B. (2005). Game Theory in Philosophy. *Topoi*, *24*(2), 197–208. doi:10.1007/s11245-005-5055-3

Chard, K., Bubendorfer, K., Caton, S., & Rana, O. F. (2012). Social cloud computing: A vision for socially motivated resource sharing. Services Computing. *IEEE Transactions on Service Computing*, *5*(4), 551-563.

Corcoran, P. (2016). The Internet of Things: Why now, and what's next? IEEE Consumer Electronics Magazine, 63-68.

Dastjerdi, A. V., & Buyya, R. (2016). Fog Computing: Helping the Internet of Things Realize Its Potential. *Computer*, *49*(8), 112–116. doi:10.1109/MC.2016.245

Du, Z., Wang, Q., Fu, X., & Liu, Q. (2012). Integrated and flexible data management for cloud social network service platform on campus. *IEEE ICCSNT*, *2012*, 1241–1244.

Ghoreyshi, S. M., Shahrabi, A., & Boutaleb, T. (2016). An Opportunistic Void Avoidance Routing Protocol for Underwater Sensor Networks. *IEEE AINA*, *2016*, 316–323.

Gibbons, R. (1992a). *A Primer in Game Theory*. Prentice Hall.

Gibbons, R. (1992b). *Game Theory for Applied Economists*. Princeton, NJ: Princeton University Press.

Han, Z., Niyato, D., Saad, W., Başar, T., & Hjørungnes, A. (2011). *Game Theory in Wireless and Communication Networks*. Cambridge University Press. doi:10.1017/CBO9780511895043

Hung, S. C., Hsu, H., Lien, S. Y., & Chen, K. C. (2015). Architecture Harmonization Between Cloud Radio Access Networks and Fog Networks. *IEEE Access*, *3*, 3019–3034. doi:10.1109/ACCESS.2015.2509638

Hwang, Y. C. & Shiau, W. C. (2012). Exploring Imagery-driven Service Framework on Social Network Service. *IEEE/ACM ASONAM'2012*, (pp. 1117-1122).

Khansalee, E., Zhao, Y., Leelarasmee, E., & Nuanyai, K. (2014). A dual-band rectifier for RF energy harvesting systems. *IEEE ECTI-CON*, *2014*, 1–4.

Kim, S. (2016). Dynamic social cloud management scheme based on transformable Stackelberg game. *EURASIP Journal on Wireless Communications and Networking*, (1), 1–9.

Kim, S. W. (2014). *Game Theory Applications in Network Design*. IGI Global. doi:10.4018/978-1-4666-6050-2

Li, L., Li, S., & Zhao, S. (2014). QoS-Aware Scheduling of Services-Oriented Internet of Things. *IEEE Transactions on Industrial Informatics*, *10*(2), 1497–1505. doi:10.1109/TII.2014.2306782

Liang, X., & Xiao, Y. (2013). Game Theory for Network Security. *IEEE Communications Surveys and Tutorials*, *15*(1), 472–486. doi:10.1109/SURV.2012.062612.00056

Lim, T. B., Lee, N. M., & Poh, B. K. (2013). Feasibility study on ambient RF energy harvesting for wireless sensor network. *IEEE IMWS-BIO*, *2013*, 1–3.

Lu, X., Wang, P., Niyato, D., Kim, D., & Han, Z. (2015). Wireless Networks With RF Energy Harvesting: A Contemporary Survey. *IEEE Communications Surveys and Tutorials*, *17*(2), 757–789. doi:10.1109/COMST.2014.2368999

Ma, K. (2011). The game analysis of regulation of the government in the Internet of Things. *IEEE AIMSEC*, *2011*, 1672–1675.

MacKenzie, A. B., & Wicker, S. B. (2001). Game theory in communications: motivation. *IEEE GLOBECOM,* (2), 821-826.

Nash, J. (1950). The Bargaining Problem.Econometrica, Econometric Society, 155-162.

Nash, J. (1953). Two-Person Cooperative Games.Econometrica, Econometric Society, 128-140.

Nef, M. A., Perlepes, L., Karagiorgou, S., Stamoulis, G. I., & Kikiras, P. K. (2012). Enabling qos in the internet of things. *CTRQ*, *2012*, 33–38.

Osborne, M. J., & Rubinstein, A. (1994). *A Course in Game Theory*. MIT Press.

Park, S., Heo, J., Kim, B., Chung, W., Wang, H., & Hong, D. (2012). Optimal mode selection for cognitive radio sensor networks with RF energy harvesting. *IEEE PIMRC*, *2012*, 2155–2159.

Park, S., Simeone, O., & Shamai, S. (2016). Joint optimization of cloud and edge processing for fog radio access networks. *IEEE ISIT*, *2016*, 315–319.

Roy, S., Ellis, C., Shiva, S., Dasgupta, D., Shandilya, V., & Wu, Q. (2010). A Survey of Game Theory as Applied to Network Security. *IEEE HICSS*, (pp. 1-10).

Rubinstein, A. (1982). Perfect equilibrium in a bargaining model. *Econometrica*, *50*(1), 97–109. doi:10.2307/1912531

Shah, P. M., Ullah, I., Khan, T., Hussain, M. S., Khan, Z. A., Qasim, U., & Javaid, N. (2016). MobiSink: Cooperative routing protocol for underwater sensor networks with sink mobility. *IEEE AINA*, *2016*, 189–197.

Sheng, X., Tang, J., Xiao, X., & Xue, G. (2014). Leveraging GPS-Less Sensing Scheduling for Green Mobile Crowd Sensing. *IEEE Internet of Things Journal, 1*(4), 328–336. doi:10.1109/JIOT.2014.2334271

Shubik, M. (1973). *Game Theory and Political Science* (Cowles Foundation Discussion Paper No.351). Department of Economics, Yale University.

Tandon, R., & Simeone, O. (2016). Harnessing cloud and edge synergies: Toward an information theory of fog radio access networks. *IEEE Communications Magazine*, *54*(8), 44–50. doi:10.1109/MCOM.2016.7537176

Wahid, A., & Kim, D. (2012). An energy efficient localization-free routing protocol for underwater wireless sensor networks. *International Journal of Distributed Sensor Networks*, *8*(4), 1–11. doi:10.1155/2012/307246

Wooldridge, M. (2012). Does Game Theory Work? *IEEE Intelligent Systems*, *27*(6), 76–80. doi:10.1109/MIS.2012.108

KEY TERMS AND DEFINITION

Cloud Computing: A type of Internet-based computing that provides shared computer processing resources and data to computers and other devices on demand.

CloudIoT: A novel paradigm where Cloud and IoT are merged together is foreseen as disruptive and as an enabler of a large number of application scenarios.

Fog Computing: Fog computing or fog networking is an architecture that uses one or more collaborative multitude of end-user clients or near-user edge devices to carry out a substantial amount of storage, communication, and control, configuration, measurement and management.

Future Internet: A general term for research activities on new architectures for the Internet.

Internet of Things: The internet of things (IoT) is the internetworking of physical devices, vehicles, buildings and other items - embedded with electronics, software, sensors, actuators, and network connectivity that enable these objects to collect and exchange data.

M2M Communications: Machine-to-machine (M2M) communications is used for automated data transmission and measurement between mechanical or electronic devices.

Quality of Service: Quality of service (QoS) is the overall performance of a telephony or computer network, particularly the performance seen by the users of the network.

Social Network Service: An online platform that is used by people to build social networks or social relations with other people who share similar personal or career interests, activities, backgrounds or real-life connections

Underwater Sensor Networks: Underwater sensor network using acoustic communications based on acoustic wireless communication in underwater environment.

Chapter 2
Cloud-Based IoT System Control Problems

ABSTRACT

Cloud computing and IoT are two very different technologies that are both already part of our life. Their adoption and use are expected to be more and more pervasive, making them important components of the Future Internet. A novel paradigm where Cloud and IoT are merged together is foreseen as disruptive and as an enabler of a large number of application scenarios. In this chapter, we focus our attention on the integration of Cloud and IoT. Reviewing the rich and articulate state of the art in this field, some issues are selected; Cloud Radio Access Network (C-RAN), Mobile Cloud IoT (MCIoT), Social Cloud (SC) and Fog Radio Access Network (F-RAN). C-RAN provides infrastructure layer services to mobile users by managing virtualized infrastructure resources. SC is a service or resource sharing framework on top of social networks, and built on the trust-based social relationships. In recent years, the idea of SC has been gaining importance because of its potential applicability. With an explosive growth of Mobile Cloud (MC) and IoT technologies, the MCIoT concept has become a new trend for the future Internet. MCIoT paradigm extends the existing facility of computing process to different mobile applications executing in mobile and portable devices. As a promising paradigm for the 5G wireless communication system, a new evolution of the cloud radio access network has been proposed, named as F-RANs. It is an advanced socially-aware mobile networking architecture to provide a high spectral and energy efficiency while alleviating backhaul burden. With the ubiquitous nature of social networks and cloud computing, IoT technologies exploit these developing new paradigms.

DOI: 10.4018/978-1-5225-1952-2.ch002

NEWS-VENDOR GAME BASED RESOURCE ALLOCATION (NGRA) SCHEME

C-RAN has been emerging as a cost-effective solution supporting huge volumes of mobile traffic in the big data era. To exploit next generation C-RAN operations, a main challenging issue is how to properly control system resources. Recently, S. Kim proposed the *News-vendor Game based Resource Allocation* (NGRA) scheme, which is a novel resource management algorithm for C-RAN systems. By employing the news-vendor game model, the NGRA scheme investigates a resource allocation problem with bargaining solutions. In dynamic C-RAN environments, this game-based resource management approach can practically adapt current system conditions while maximizing the expected payoff.

Development Motivation

Modern computation and communication systems operate in a new and dynamic world, characterized by continual changes in the environment and performance requirements that must be satisfied. Dynamic system changes occur without warning and in an unpredictable manner, which are outside the control of traditional operation approaches (Addis, 2013). At the same time, popularity of mobile devices and related applications in various fields are increasing significantly in everyday life. Furthermore, applications become more and more complex, Quality of Service (QoS) sensitive and computation intensive to perform on mobile system. Therefore, new solution concepts need to be developed that manage the computation and communication systems in a dynamically adaptive manner while continuously ensuring different application services (Addis, 2013; Htikie, 2013).

Cloud Radio Access Network (C-RAN) is a new system architecture for the future mobile network infrastructure. It is a centralized, cloud computing based new radio access network to support future wireless communication standards. C-RAN can be implemented based on the concept of virtualization. Usually, virtualization is an enabling technology that allows sharing of the same physical machine by multiple end-user applications with QoS guarantees. Therefore, it helps to reduce costs while improving a higher utilization of the physical resources (Addis, 2013; Zhu, 2014).

No one may deny the advantages of C-RAN services via virtualization technologies. However, there are some problems that need to be addressed. Most of all, next generation C-RAN systems should take into account QoS guarantees while maximizing resource efficiency. However, because of the scarcity of system resource, it is difficult to satisfy simultaneously these conflicting requirements. For this reason, the most critical issue for the next generation C-RAN system is to develop effective resource allocation algorithms (Vakilinia, 2014). But, despite flexibility and great potential applicability, resource allocation problem in C-RAN has received scarce attention as of today.

To design a resource allocation algorithm in C-RAN systems, it is necessary to study a strategic decision making process. Under widely dynamic C-RAN conditions, system agents can be assumed as intelligent rational decision-makers, and they select a best-response strategy to maximize their expected utility with other agents. This situation is well-suited for the game theory. News-vendor game (Malakooti, 2014; William, 2009) is a mathematical game model in operations management and applied economics used to determine optimal inventory levels. Typically, it is characterized by fixed prices and uncertain demand for a perishable product. Therefore, this model can represent a situation faced by a newspaper vendor who must decide how many copies of the day's paper to stock in the face of uncertain demand and knowing that unsold copies will be worthless at the end of the day. The original concept of newsvendor game appeared to date from 1888 where F. Edgeworth used the central limit theorem to determine the optimal cash reserves to satisfy random withdrawals from depositors. The modern formulation dates from the 1951 paper in Econometrica by K. Arrow, T. Harris, and J. Marshak (Arrow, 1951).

Motivated by the aforementioned discussion, the *NGRA* scheme was developed. The main goal of the *NGRA* scheme is to maximize resource efficiency while providing QoS guarantees. In dynamically changing C-RAN environments, the game process in the *NGRA* scheme is divided two stages; the competitive stage and the bargaining stage. At the competitive stage, system resource is allocated in a non-cooperative game manner. Therefore, cloud server controls dynamically the total service request by adjusting the price. When the total service request is larger than the system capacity with the maximum price, system resource can not be distributed effectively in a non-cooperative manner. To effectively handle this case, the *NGRA* scheme adopts a bargaining-based approach. At the bargaining stage, the *NGRA* scheme re-distributes the system resource on the basis of combined bargaining solution.

Cloud Radio Access Network Architecture

In C-RAN systems, there are multiple Cloud Providers (*CPs*), which can generate more revenue from the sharing of available resources. *CPs* have their system resources, such as CPU core, memory, storage, network bandwidth, etc. To ensure the optimal usage of cloud resources, baseband processing is centralized in a Virtualized Baseband units Pool (*VBP*). The *VBP* can be shared by different *CPs* and multiple Base Stations (*BSs*). Therefore, the *VBP* is in a unique position as a cloud brokering between the *BSs* and the *CPs* for cloud services while increasing resource efficiency and system throughput (Checko, 2015).

In the C-RAN architecture, the *BS* covers a small area, and communicates with the Mobile Users (*MUs*) through wireless links. The *BSs* provide the managed connectivity and offer flexibility in real-time demands. To improve *C-RAN* system efficiency, *CPs* can offer their available resources to *BSs* through the *VBP*, and *BSs* can provide services to *MUs* based on their obtained resources. Without loss of generality, each *BS* is assumed to acts as a virtual machine, and *MUs*' applications are executed through the virtualization technology. The general architecture of hierarchical C-RAN system is shown in Figure 1.

Figure 1. Hierarchical C-RAN system structure

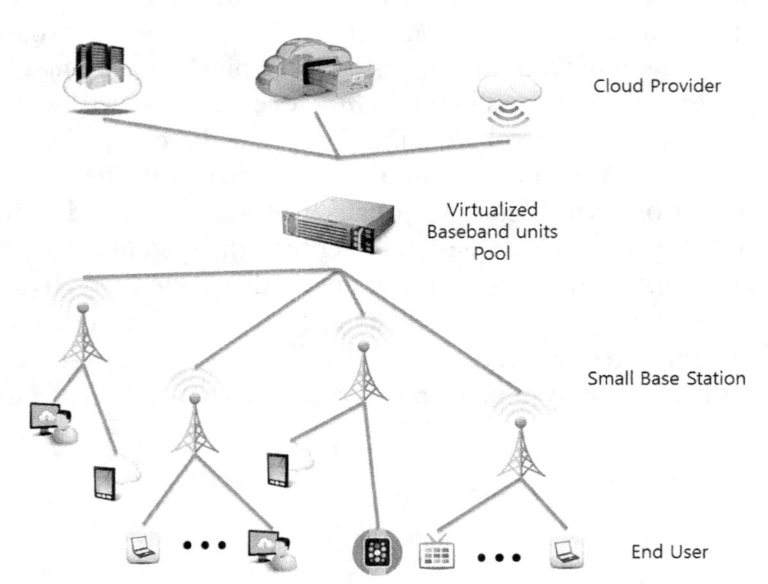

News-Vendor Game Models at the Competitive Stage

The *NGRA* scheme introduces the news-vendor game model ($\mathbb{G}$) for C-RAN systems. $\mathbb{G}$ is a tuple $\left(V, \mathbb{N}, \left(\boldsymbol{S}_i\right)_{i\in\mathbb{N}}, \left(U_i\right)_{i\in N}, T\right)$ at each time period t of game-play.

- V is the total amount of available cloud resource in the *VBP*.
- $\mathbb{N}$ is the finite set of players $\mathbb{N} = \left\{b_0, b_1, \ldots, b_n\right\}$ where b_0 is the *VBP* and $b_{i,1\leq i\leq n}$ represents the i th *BS*.
- $\boldsymbol{S}_i$ is the set of strategies with the player i. If the player i is the *VBP*, i.e., i = 0, a strategy set can be defined as resource prices. If the player i is a *BS*, i.e., $1 \leq i \leq n$, the strategy set is defined as the amount of requested resource.
- The U_i is the payoff received by the player i. Traditionally, the payoff is determined as the obtained outcome minus the cost to obtain that outcome. For simplicity, the outcome is represented in a general form of *log* function.
- The T is a time period. The $\mathbb{G}$ is repeated $t \in T < \infty$ time periods with competitive and cooperative manner.

To understand the behavior of self-regarding system agents, game models have some attractive features. As a kind of game model, news-vendor game was initially developed for the classical, single-period newsboy problem (Wen, 2009). In the *NGRA* scheme, the traditional news-vendor game is extended as a two-stage repeated game. Initially, the *VBP* dynamically adjusts the price of resource unit, and *BSs* request cloud resources to maximize their payoffs. In this stage, resource allocation procedure is formulated as a non-cooperative game model. If service requests from *BSs* is more than the capacity of the *VBP*, the cloud resource is re-distributed adaptively according to the combined bargaining solution. In this stage, resource allocation procedure is formulated as a bargaining game approach. By a sophisticated combination of these two different game approaches, the *NGRA* scheme attempts to approximate a well-balanced system performance among conflicting requirements.

At the competitive stage, the strategy set for the *VBP* ($\boldsymbol{S}_0$), i.e., available price levels for a resource unit, is assumed as below.

$$S_0 = \left\{p^t \mid p^t \in \left[p_{\min}, p_{\max}\right]\right\} \tag{1}$$

where p^t is the price at time t. The $p_{\min}, p_{\max}$ are the pre-defined minimum and maximum price levels, respectively. From the viewpoint of *VBP*, the $p_{\max}$ is good to maximize its profit. From the viewpoint of *BSs*, $p_{\min}$ is good to maximize their payoff. The actual price at time t (p^t) is dynamically decided according to the current system conditions. In this model, the p^t is determined as the weighted sum of $p_{\min}$ and $p_{\max}$.

$$p^t = \omega \times p_{\max} + \left(1 - \omega\right) \times p_{\min} \tag{2}$$

where ω is a weighted factor for the both prices. Under diverse system environments, the value of ω should be modified dynamically. In the *NGRA* scheme, *Rubinstein-Stahl* model is adopted to adjust the ω value. *Rubinstein-Stahl* model was proposed as a solution to the problem when two players were negotiating the division of the benefit (Xie, 2010). Therefore, players negotiated with each other by proposing offers alternately. After several rounds of offer and count-offers, players finally come to an agreement. In *Rubinstein-Stahl* model, there exists a unique solution for this negotiation process (Xie, 2010). The *NGRA* scheme assumes that the *Rubinstein–Stahl* model's equilibrium point is obtained through negotiation between the *VBP* stance and *BSs*' stance.

In the *Rubinstein-Stahl* model, the *VBP* is assumed as a supplier, and all *BSs* are assumed as a single customer. Two players, i.e., supplier and customer, have their own bargaining power (δ). The division proportion of the benefits can be obtained according to the negotiation power, which can be computed at each player individually. A more negotiation power player benefits more from the negotiation process. Players negotiate with each other by proposing offers alternately. After several rounds of negotiation, they finally reach an agreement as following (Park, 2015; Zhao, 2002).

$$\left(x_s^*, x_c^*\right) = \begin{cases} \left(\dfrac{1-\delta_c}{1-\delta_s \times \delta_c}, \dfrac{\delta_c \times \left(1-\delta_c\right)}{1-\delta_s \times \delta_c}\right) & \textit{if supplier offers first} \\ \left(\dfrac{\delta_s \times \left(1-\delta_c\right)}{1-\delta_s \times \delta_c}, \dfrac{1-\delta_s}{1-\delta_s \times \delta_c}\right) & \textit{if customer offers first} \end{cases}.$$

s.t.,

$$\left(x_s^*, x_c^*\right) \in R^2 : x_s^* + x_c^* = 1, x_s^* \geq 0, x_c^* \geq 0 \text{ and } 0 \leq \delta_s, \delta_c \leq 1 \quad (3)$$

where x_s^* and x_c^* are final dividends for supplier and customer, respectively. δ_s and δ_c be the supplier and consumer patience factor. Lower δ (or higher δ) value means lower patience (or more patience). In the *Rubinstein-Stahl* negotiation model, the patience factor strongly affects the negotiation process; the more patience has, the more payoff attains. From a common-sense standpoint, consumers should know the current price as early as possible for the effective service continuity. Under this situation, they lack patience in bargaining. For this reason, the *NGRA* scheme represents the consumer's patience as a monotonous time decreasing function. According to the inverse effect of reciprocal relationship, the supplier's patience is defined vice versa. Therefore, the consumer's patience (δ_c^t) and supplier's patience (δ_s^t) at t th round of negotiation process are defined as follow (Park, 2015; Xie, 2010).

$$\delta_c^t = 1 - \left(\frac{e^{\xi^t} - e^{-\xi^t}}{e^{\xi^t} + e^{-\xi^t}}\right) \text{ and } \delta_s^t = \left(\frac{e^{\xi^t} - e^{-\xi^t}}{e^{\xi^t} + e^{-\xi^t}}\right)$$

s.t.,

$$\frac{d\delta_c^t}{dt} < 0,\ \frac{d\delta_s^t}{dt} > 0,\ \delta_c^0, \delta_s^\infty = 1, \text{ and } \delta_c^\infty, \delta_s^0 = 0 \quad (4)$$

where ξ^t is the patience coefficient at t th round. For the ideal C-RAN system management, the *NGRA* scheme dynamically adjusts the ξ^t value. When the requested service increases (*or* decreases), the price (p) should increases (*or* decreases) to maximize the resource efficiency. To implement this mechanism, the value of ξ^t is defined as the ratio of the current cloud workload to the total system capacity.

$$\xi^t = 1 - \left(\frac{T_Q - C_Q}{T_Q}\right), s.t., 0 < \xi^t < 1 \quad (5)$$

where T_Q and C_Q are the total cloud capacity and the current cloud workload, respectively. If the gap between T_Q and C_Q is larger, the ξ^t decreases using

(5) and δ_c $(or\delta_s)$ increases (*or* decreases), simultaneously. Therefore, according to the ξ^t value, the values of δ_c and δ_s are adjusted adaptively. On the basis of obtained δ_c and δ_s values, the *NGRA* scheme can get the weighted factor (ω). In a realistic negotiation scenario, a supplier offers the price first. Therefore, the values of ω and $(1-\omega)$ are obtained according to (3).

$$\omega = \frac{1-\delta_c}{1-\delta_s \times \delta_c} \text{ and } (1-\omega) = \frac{\delta_c \times \left(1-\delta_s\right)}{1-\delta_s \times \delta_c} \tag{6}$$

Finally, the price (p) in the competitive game stage is obtained based on the equation (2). When the p is high, cloud service requests are reduced with unsatisfactory payoffs, and vice versa. Therefore, at the competitive stage, the *VBP* can control dynamically the total service request by adjusting the price according to (2)-(6).

News-Vendor Game Models at the Cooperative Stage

When the current cloud workload is controllable through the above price strategy, the news-vender game can be operated only in the competitive game stage. However, in an overloaded situation, i.e., the cloud resource is not sufficient to support all service requests, the cooperative game stage is started. In recent years, cooperative approaches derived from game theory have been widely used for efficient resource allocation problems. The most popular approaches are the *Nash Bargaining Solution* (NBS) and the *Kalai-Smorodinsky bargaining solution* (KSBS) (Kim, 2010; Kim, 2014). Because of their appealing properties, the basic concept of NBS and KSBS has become an interesting research topic in economics, political science, sociology, psychology, biology, and so on (Kim, 2014).

Based on the traditional game theory, the Nash bargaining solution can be formulated as follows (Kim, 2014).

$$\prod_i (u_i^* - d_i) = \max_{u_i \in \mathbb{S}} \prod_i \left(u_i - d_i\right), where\ u_i^* \in \mathbb{S}\ and\ d_i \in \boldsymbol{d} \tag{7}$$

where $\mathbb{S} = \{(u_1, \ldots u_n)\} \subset \mathbb{R}^n$ is a jointly feasible utility solution set, and a disagreement point (***d***) is an action vector $\boldsymbol{d} = (d_1,,, d_n) \in \mathbb{S}$ that is expected to be the result if players cannot reach an agreement. In the game theory terminology, an outcome vector $< u_1^*, u_2^* ,,, u_n^* >$ is a unique and fair-efficient solution, called the NBS that fulfills the Nash axioms (Kim, 2014).

1. **Individual Rationality:** NBS should be better off than the disagreement point. Therefore, no player is worse off than if the agreement fails. Formally, $u_i^* \geq d_i$ for all player i.
2. **Feasibility:** NBS is reasonable under the circumstances. That is, $\boldsymbol{U}^* \in \mathbb{S}$.
3. **Pareto Optimality:** NBS gives the maximum payoff to the players. Therefore, if there exists a solution $\boldsymbol{U}^* = \left(u_1^*..u_i^*..u_n^*\right)$, it shall be Pareto optimal.
4. **Invariance with Respect to Utility Transformations:** A utility function specifies a player's preferences. Therefore, different utility functions can be used to model the same preferences. However, the final outcome should not depend on which of these equivalent utility representations is used. In other words, for any linear scale transformation of the function ψ, $\psi\left(F\left(\mathbb{S},d\right)\right) = F\left(\psi\left(\mathbb{S}\right),\psi\left(d\right)\right)$. This axiom is also called *Independence of Linear Transformations or scale covariance.*
5. **Independence of Irrelevant Alternatives:** The solution should be independent of irrelevant alternatives. In other words, a reasonable outcome will be feasible after some payoff sets have been removed. If $\boldsymbol{U}^*$ is a bargaining solution for a bargaining set $\mathbb{S}$ then for any subset $\mathbb{S}'$ of $\mathbb{S}$ containing $\boldsymbol{U}^*$, $\boldsymbol{U}^*$ continues to be a bargaining solution. Formally, if $\boldsymbol{U}^* \in \mathbb{S}' \subset \mathbb{S}$ and $U^* = F\left(\mathbb{S},d\right)$, then $\boldsymbol{U}^* = F\left(\mathbb{S}',\mathrm{d}\right)$.
6. **Symmetry:** Symmetry means that if the players' utilities are exactly the same, they should get symmetric payoffs, i.e., equal payoffs. Therefore, payoff should not discriminate between the identities of the players, but only depend on utility functions. For example, if $\mathbb{S}$ is invariant under all exchanges of users, $F_i\left(\mathbb{S},d\right) = F_j\left(\mathbb{S},d\right)$ for all possible players i and j.

Even though the NBS can provide a unique and fair Pareto optimal solution, Nash axioms do not always characterize the situations we encounter in reality. In particular, the *Independence of irrelevant alternatives* has been the source of considerable contention. When a feasible solution set is modified, NBS is unconcerned about a relative fairness. Therefore, the dilemma is an insensitivity to utility translations. In some cases, the outcome of the bargaining process may be the result of reciprocal equality. Therefore, during the 1950-1980s, extensive research had been done to replace the axiom, *Independence of irrelevant alternatives* (Zehavi, 2009).

KSBS is an alternative approach to the bargaining problem proposed by Kalai and Smorodinsky (Kim, 2010; Kim, 2014). While Nash's solution requires the solution to be independence when irrelevant alternatives are modified, the KSBS relaxed this condition. Therefore, Kalai and Smorodinsky replaced the axiom of *Independence of irrelevant alternatives* by *individual monotonicity*. Under *individual monotonicity* condition, if the feasible set is changed in favor of one of the players, this player should not end up losing because of this change (Zehavi, 2009). More formally, *individual monotonicity* axiom is defined as (Kim, 2010; Kim, 2011).

7. **Individual Monotonicity:** A bargaining situation $\left(\mathbb{W}, d\right)$ is better than $\left(\mathbb{S}, d\right)$. if and only if

$$\sup\left\{u_i : \left\{\left(u_1,,,u_n\right)\right\} \in \mathbb{W}\right\} \geq \sup\left\{u_i : \left\{\left(u_1,,,u_n\right)\right\} \in \mathbb{S}\right\}$$

where $1 \leq i \leq n$. A solution function F is individually monotonic for a player if whenever $\left(\mathbb{W}, d\right)$ is better than $\left(\mathbb{S}, d\right)$, then $F\left(\mathbb{W}, d\right) > F\left(\mathbb{S}, d\right)$. F is individually monotonic if the same property holds for all players.

KSBS is a unique solution satisfying the axioms (1)-(4) and (6)-(7). Mathematically, it is defined as

$$\frac{\sup\{u_1\} - d_1}{I_1^* - d_1} = \ldots = \frac{\sup\{u_i\} - d_i}{I_i^* - d_i} = \ldots = \frac{\sup\{u_n\} - d_n}{I_n^* - d_n}$$

s.t.,

$$\sup\{u_i\} = \sup\left\{u_i : \left\{\left(u_1,,,u_n\right)\right\} \in \mathbb{W}\right\},\ I_i^* = \max\left\{u_i : u_i \in \mathbb{S}\right\} \text{ and } 1 \leq i \leq n \tag{8}$$

where I_i^* is the ideal point of player i. Therefore, players choose the best outcome subject to the condition that their proportional part of the excess over the disagreement is relative to the proportion of the excess of their ideal gains (Kim, 2010; Kim, 2011; Zehavi, 2009).

The *NGRA* scheme develops a new bargaining solution by combining the axioms of *Independence of irrelevant alternatives* and *Individual monotonicity*. To implement this solution, the main issue is how to trade-off between different principles, which can be tackled by cooperative games with transferable utility. The combined bargaining solution (u^{α}) of the NBS and KSBS is such that:

$$U_i^{\alpha} = \alpha \times u_i^{NBS} + \left(1-\alpha\right) \times u_i^{KSBS}$$

$$U_i^{\alpha} \in \left\{U_1^{\alpha}..U_i^{\alpha}..U_n^{\alpha}\right\} \text{and } 1 \le i \le n \tag{9}$$

where α is a control parameter to relatively emphasize the principle of *Independence of irrelevant alternatives* or *Individual monotonicity*. The major feature of *Individual monotonicity* is that increasing the bargaining set size in a direction favorable to a specific player always benefits that player. Therefore, when the bargaining set size of each player is huge different, this feature can keep the relative fairness among players.

The *NGRA* scheme adaptively adjusts the α value in an online decision manner. In dynamic C-RAN environments, a fixed value of α cannot effectively adapt to the changing conditions. When the normalized difference of the bargaining set size is high, the *NGRA* scheme should strongly depend on the axiom of *Individual monotonicity*. In this case, a lower value of α is more suitable. When the normalized difference of the bargaining set size is nearly the same, the *NGRA* scheme can put more emphasis on the axiom of *Independence of irrelevant alternatives*. In this case, a higher value of α is more desirable. Based on this consideration, the value of α is dynamically adjusted according to the current ratio of bargaining set difference.

$$\alpha = \frac{\min_{i,j \in \mathbb{N}}\left(\left|I_i^* - I_j^*\right|\right)}{\max_{i,j \in \mathbb{N}}\left(\left|I_i^* - I_j^*\right|\right)} s.t., 1 \le i, j \le n \tag{10}$$

With the dynamic adaptation of α value, the *NGRA* scheme can be more responsive to current C-RAN conditions. Finally, the set of resource allocation for BSs at time t, denoted by $\mathbb{R}$, is calculated as

$$\mathbb{R} = \mathcal{R}_i \mid \left\{\mathcal{R}_1..\mathcal{R}_i..\mathcal{R}_n\right\} \underline{\underline{def}} \mathcal{R}_i = \left(\frac{U_i^{\alpha}}{\sum_{i=1}^{n} U_i^{\alpha}} \right) \times \mathfrak{R}^t \tag{11}$$

where $\mathfrak{R}^t$ is the available cloud resource at time t. The solution set $\mathbb{R}$ is a possible outcome of the combined bargaining process; $\mathbb{R}$ is adaptively obtained in the cooperative trade-off area.

The Main Steps of the *NGRA* Scheme

In the *NGRA* scheme, both non-cooperative and cooperative game models have been applied to the cloud resource allocation process. This two-stage approach suggests that a judicious mixture of collaboration and competition is advantageous in dynamic C-RAN environments. The main steps of the *NGRA* scheme are given next.

Step 1: At the initial time, the price (p) is set to the initial value, and *BSs* request their cloud service to maximize their payoffs in a non-cooperative game approach.

Step 2: At each game period, the p is decided according to the *Rubinstein-Stahl* model. In the basis of (2)-(6), the p is dynamically adjusted by taking into account the current cloud workload (ξ).

Step 3: After the p decision, δ, ω and ξ values are modified periodically using (4),(5) and (6).

Step 4: When the cloud services are congested at the competitive stage, it is impossible to control the resource allocation through price control strategy. At this time, the bargaining stage is started.

Step 5: At each game period, NBS and KSBS are obtained using (7) and (8). At the same time, the control parameter α is adjusted dynamically using (10).

Step 6: The combined bargaining solution ($U^{\alpha}_{i,1\leq i\leq n}$) of the NBS and KSBS are obtained based on (9), and the set of resource allocation for *BSs* ($\mathbb{R}$) is finally calculated according to (11).

Step 7: Under widely diverse C-RAN environments, the *VBP* and *BSs* are self-monitoring constantly for the next news-vendor game process; proceed to Step 2.

Summary

As a new model of distributed computing, all kinds of distributed resources are virtualized to establish a shared resource pool through C-RAN systems. C-RAN solution enables dynamic on-demand response, combining collaborative radio and real-time cloud infrastructure while providing convenient and configurable resources. Therefore, dynamic and efficient mechanism for rapidly scaling cloud resources is becoming a hot spot in research areas. The NGRA scheme is a novel resource allocation scheme based on the newsvender game model. The main goal of the NGRA scheme is to maximize system performance while ensuring service QoS. To satisfy this goal, the NGRA scheme develops a two-stage game mechanism. The important feature of this approach is its adaptability, flexibility and responsiveness to current C-RAN conditions.

THE HIERARCHICAL GAME BASED RESOURCE SHARING (HGRS) SCHEME

Over the past decade, wireless applications have experienced tremendous growth, and this growth is likely to multiply in the near future. To cope with expected drastic data traffic growth, Cloud computing based new Radio Access Network (*C-RAN*) has been proposed for next generation cellular networks. It is considered as a cost efficient way of meeting high resource demand of future wireless access networks. The Hierarchical Game based Resource Sharing (HGRS) scheme is a novel resource sharing scheme for future *C-RAN* systems. Based on the Indian buffet game, the *Hierarchical Game based Resource Sharing* (HGRS) scheme formulates the *C-RAN* resource allocation problem as a two-level game model, and finds an effective solution according to the coopetition approach.

Development Motivation

In recent years, the Radio Access Network (*RAN*) is commonly used to support the exponential growth of mobile communications. Conceptually, *RAN* resides among network devices such as a mobile phone, a computer, or any remotely controlled machine, and provides connections with core networks. However, traditional *RAN* architecture has been faced with a number of challenges. First, a highly loaded Base Station (*BS*) cannot share processing power with other idle or less loaded *BSs*; it results in a poor resource utilization.

Second, a *BS* equipment serves only radio frequency channels in each physical cell, where *BS*'s resources cannot be shared with other *BSs* in different cells. Finally, *BSs* built on proprietary hardware cannot have a flexibility to upgrade radio networks (Htike, 2013; Sigwele, 2014). To overcome these problems, Cloud computing based Radio Access Network (*C-RAN*) is widely considered as a promising paradigm, which can bridge the gap between the wireless communication demands of end-users and the capacity of radio access networks (Sigwele, 2014; Zhu, 2014).

In 2013, C. Jiang introduced the fundamental notion of Indian buffet game to study how game players make multiple concurrent selections under uncertain system states (Jiang, 2013). Specifically, Indian buffet game model can reveal how players learn the uncertainty through social learning and make optimal decisions to maximize their own expected utilities by considering negative network externality (Jiang, 2015). This game model is well-suited for the *C-RAN* resource sharing problem. Motivated by the above discussion, the HGRS scheme is developed based on the Indian game model. The key feature of HGRS scheme is to develop a decentralized mechanism according to the two-level coopetition approach. The term 'coopetition' is a neologism coined to describe cooperative competition. Therefore, coopetition is defined as the phenomenon that differs from competition or cooperation, and stresses two faces, i.e., cooperation and competition, of one relationship in the same situation (Kim, 2015). In the HGRS scheme, the game model consists of two-levels; upper and lower Indian buffet games. At the upper-level game, cloud resources are shared in a cooperative manner. At the lower-level game, allocated resources are distributed in a non-cooperative manner. Based on the hierarchical interconnection of two game model, control decisions can cause cascade interactions to reach a mutually satisfactory solution.

Usually, different *C-RAN* agents may pursue different interests, and act individually to maximize their own profits. This self-organizing feature can add autonomics into *C-RAN* systems and help to ease the heavy burden of complex centralized control algorithms. Based on the recursive best response algorithm, the HGRS scheme draws on the concept of a learning perspective and investigates some of the reasons and probable lines for justifying each system agent's behaviors. The dynamics of the interactive feedback learning mechanism can allow control decisions to be dynamically adjustable. In addition, by employing the coopetition approach, control decisions are mutually dependent on each other to resolve conflicting performance criteria.

Indian Buffet Game Model for *C-RAN* Systems

The HGRS scheme considers a *C-RAN* architecture with one Virtualized Baseband units Pool (*VBP*), 10 Small Base Stations (*SBSs*) and 100 Mobile Users (*MUs*), and system resources, which are the computing capacities of CPU, memory, storage and bandwidth. These resources can be used by the *MUs* through the *VBP* to gain more revenue. *CPs* cooperate to form a logical pool of computing resources to support *MUs'* applications. Each *MU* application service has its own application type, and requires different resources requirements.

Let us consider an Indian buffet restaurant which provides m dishes denoted by $d_1, d_2, \ldots, d_m$. Each dish can be shared among multiple guests. Each guest can select sequentially multiple dishes to get different meals. The utility of each dish can be interpreted as the deliciousness and quantity. All guests are rational in the sense that they will select dishes which can maximize their own satisfactions. In such a case, the multiple dish-selection problem can be formulated to be a non-cooperative game, called Indian buffet game. In the traditional Indian buffet game, the main goal is to study how guests in a buffet restaurant learn the uncertain dishes' states and make multiple concurrent decisions by not only considering the current utility, but also taking into account the influence of subsequent players' decisions (Kim, 2014; Jiang, 2015).

During the *C-RAN* system operations, system agents should make decisions individually. In this situation, a main issue for each agent is how to perform well by considering the mutual-interaction relationship and dynamically adjust their decisions to maximize their own profits. The HGRS scheme develops a new *C-RAN* system resource sharing algorithm based on the Indian buffet game model. In the HGRS scheme, the dynamic operation of *VBP*, *SBSs* and *MUs* is formulated as a two-level Indian buffet game. At the first stage, the *VBP* and *SBSs* play the upper-level Indian buffet game; the *VBP* distribute the available resources to each *SBS* by using a cooperative manner. At the second stage, multiple *MUs* decide to purchase the resource from their corresponding *SBS* by employing a non-cooperative manner. Based on this hierarchical coopetition approach, the HGRS scheme assumes that all game players (*VBP*, *SBSs* and *MUs*) are rational and independent of gaining the profit as much as possible. Therefore, for the implementation practicality, the HGRS scheme is designed in an entirely distributed and self-organizing interactive fashion.

Mathematically, the upper-level Indian buffet game ($\mathbb{G}^U$) can be defined as

$$\mathbb{G}^L = \left\{\mathbb{P}, \left\{\mathcal{L}_i\right\}_{i\in\mathbb{P}}, \left\{\mathcal{T}_i\right\}_{i\in\mathbb{P}}, \left\{U_i\right\}_{i\in\mathbb{P}}, T\right\}$$

at each time period t of gameplay.

- $\mathbb{N}$ is the finite set of players $\mathbb{N} = \left\{\mathcal{C}, \mathcal{B}\right\}$ where $\mathcal{C}$ = {*VBP*} represents one *VBP* and $\mathcal{B} = \left\{b_1, \ldots, b_n\right\}$ is a set of multiple *SBSs*, which are assumed as guests in the upper-level Indian restaurant.
- $\mathbb{D}$ is the finite set of resources $\mathbb{D} = \{d_1, d_2, \ldots, d_l\}$ in the *VBP*. Elements in $\mathbb{D}$ metaphorically represent different dishes on the buffet table in the upper-level Indian restaurant.
- $\boldsymbol{S}_i$ is the set of strategies with the player i. If the player *i* is the *VBP*, i.e., $i \in \mathcal{C}$, a strategy set can be defined as $\boldsymbol{S}_i = \{\delta_i^1, \delta_i^2, \ldots \delta_i^l\}$ where δ_i^k is the distribution status of kth resource, i.e., $1 \leq k \leq l$. If the player *i* is a *SBS*, i.e., $i \in \mathcal{B}$, the player i can request multiple resources. Therefore, the strategy set can be defined as a combination of requested resources

$$\boldsymbol{S}_i = \{\varnothing, \{d_i^1\left(\mathcal{I}_i^1\right)\}, \{d_i^1\left(\mathcal{I}_i^1\right), d_i^2\left(\mathcal{I}_i^2\right)\}, \ldots, \{d_i^1\left(\mathcal{I}_i^1\right), d_i^2\left(\mathcal{I}_i^2\right), \ldots, d_i^l\left(\mathcal{I}_i^l\right)\}\}$$

where $\mathcal{I}_i^k$ is the player i's requested amount for the kth resource; each player's strategy set is finite with 2^l elements.

- The U_i is the payoff received by the player i. If the player *i* is the *VBP*, i.e., $i \in \mathcal{C}$, it is the total profit obtained from the resource distribution for *SBSs*. If the player *i* is a *SBS*, i.e., $i \in \mathcal{B}$, the payoff is determined as the outcomes of the distributed resources minus the cost of corresponding resources.
- The T is a time period. The $\mathbb{G}^U$ is repeated $t \in T < \infty$ time periods with imperfect information.

Based on the distributed resources, *SBSs* are responsible to support *MUs*' services while ensuring the required Quality of Service (QoS). Usually, *SBSs* deploy sparsely with each other to avoid mutual interference, and are operated in a time-slotted manner. To formulate interactions between *SBSs* and *MUs*, the lower-level Indian buffet game ($\mathbb{G}^L$) can be defined as

$$\mathbb{G}^L = \left\{\mathbb{P}, \left\{\mathcal{L}_i\right\}_{i\in\mathbb{P}}, \left\{\mathcal{T}_i\right\}_{i\in\mathbb{P}}, \left\{U_i\right\}_{i\in\mathbb{P}}, T\right\}$$

at each time period t of gameplay.

- $\mathbb{P}$ is the finite set of players $\mathbb{P} = \left\{\mathcal{B}, \mathcal{X}\right\}$ where $\mathcal{B} = \left\{b_1, \ldots, b_n\right\}$ is a set of multiple *SBSs*, and $\mathcal{X} = \left\{x_1, \ldots, x_m\right\}$ is a set of multiple *MUs*, which are assumed guests in the lower-level Indian restaurant.
- $\mathcal{L}_i = \{\mathcal{R}_i^1, \mathcal{R}_i^2, \ldots, \mathcal{R}_i^l\}$ is the finite set of the player i's resources, i.e., $i \in \mathcal{B}$. Elements in $\mathcal{L}_i$ metaphorically represent different dishes on the buffet table in the ith lower-level Indian restaurant; there are total n lower-level Indian restaurants.
- $\mathcal{T}_i$ is the set of strategies with the player i. If the player i is a *SBS*, i.e., $i \in \mathcal{B}$, the strategy set can be defined as $\mathcal{T}_i = \{\lambda_i^1, \lambda_i^2, \ldots \lambda_i^l\}$ where λ_i^k is the price of kth resource in the ith*SBS*. If the player *i* is a *MU*, i.e., $i \in \mathcal{X}$, the player i can request multiple resources. Therefore, the strategy set can be defined as a combination of requested resources

$$\mathcal{T}_i = \{\varnothing, \{\mathcal{R}_i^1(\xi_i^1)\}, \{\mathcal{R}_i^1(\xi_i^1), \mathcal{R}_i^2(\xi_i^2)\}, \ldots, \{\mathcal{R}_i^1(\xi_i^1), \mathcal{R}_i^2(\xi_i^2), \ldots, \mathcal{R}_i^l(\xi_i^l)\}\}$$

where ξ_i^k is the *MU* i's request amount for the kth resource.

- The U_i is the payoff received by the player i. If the player i is a *SBS*, i.e., $i \in \mathcal{B}$, it is the total profit obtained from the resource allocation for *MUs*. If the player i is a *MU*, i.e., $i \in \mathcal{X}$, the payoff is determined as the outcomes of the allocated resources minus the cost of corresponding resources.
- The T is a time period. The $\mathbb{G}^L$ is repeated $t \in T < \infty$ time periods with imperfect information.

C-RAN Resource Sharing in Upper Indian Buffet Game

First, the upper-level Indian buffet game in the HGRS scheme is addressed. In *C-RAN* systems, there are multiple resource types, and multiple *SBSs* requests different resources to the *VBP*. The HGRS scheme mainly considered four resource types: CPU, memory, storage, and network bandwidth. Let $\mathbb{D}$ denote a set of resources in the *VBP*, $\mathbb{D} = \{d_1 =$ CPU, $d_2 =$ memory, $d_3 =$ storage, $d_4 =$ bandwidth$\}$ where each d represents the available amount of corresponding resource. Virtualization technology is used to collect these resources from *CPs*, and they are dynamically shared among *SBSs*. In the upper-level Indian buffet game, there are one *VBP* and *n SBSs*. The *VBP* is responsible for the cloud resource control, and distributes resources over multiple *SBSs*. Each *SBS* is deployed for each micro cell, and covers relatively a small area. In general, *SBSs* are situated around high traffic-density hot spots to support QoS ensured applications. To get an effective solution for the upper-level Indian game, the HGRS scheme focused on the basic concept of the *Shapley Value* (*SV*). It is a well-known solution idea for ensuring an equitable division, i.e., the fairest allocation, of collectively gained profits among the several collaborative players (Kim, 2014).

When the requested amount of kth resource (∂_i^k, $1 \le k \le 4$) of the the ith *SBS* (SBS_i) is less than the distributed resource ($\mathcal{A}_i^k$), i.e., $\partial_i^k < \mathcal{A}_i^k$, the SBS_i can waste this excess resource, and the property loss is estimated based on the resource unit price ($U\_\mathcal{P}_i^k$). $U\_\mathcal{P}_i^k$ value is adaptively adjusted in the lower-level Indian buffet game. In this case, the value function ($v(SBS_i)$) of the SBS_i becomes

$$v(SBS_i) = -U\_\mathcal{P}_i^k \times \left(\mathcal{A}_i^k - \partial_i^k\right).$$

Conversely, if $\partial_i^k > \mathcal{A}_i^k$, the deficient resource amount $\left(\partial_i^k - \mathcal{A}_i^k\right)$ is needed in the SBS_i. Therefore, the value function becomes

$$v(SBS_i) = U\_\mathcal{P}_i^k \times \left(\partial_i^k - \mathcal{A}_i^k\right).$$

The HGRS scheme assumes that

$$\mathbb{N} = \{\ \mathcal{C} = \{VBP\} \cup \mathcal{B} = \{b_1, \ldots, b_n\}\}$$

is a set of upper game players and $v(\cdot)$ is a real valued function defined on all subsets of $\mathcal{B}$ satisfying $v(\varnothing) = 0$. Therefore, in the game model, a non-empty subset ($\boldsymbol{c}$) of $\mathcal{B}$ is called a *coalition.* A set of games with a finite number of players is denoted by “ . Given a game $(\mathcal{B}, v(\cdot)) \in$ “ , let $\mathbb{C}^k$ be a *coalition structure* of $\mathcal{B}$ for the kth resource. In particular, $\mathbb{C}^k = \{\boldsymbol{c}_1^k, \ldots, \boldsymbol{c}_j^k\}$ is a partition of $\mathcal{B}$, that is, $\boldsymbol{c}_f^k \cap \boldsymbol{c}_h^k = \varnothing$ for $f \neq h$ and $\bigcup_{t=1}^{j} \boldsymbol{c}_t^k = \mathcal{B}$.

Let θ be an order on $\mathcal{B}$, that is, θ is a bijection on $\mathcal{B}$. A set of all the orders on $\mathcal{B}$ is denoted by ˜ $(\mathcal{B})$ (Kamijo, 2009; Lee, 2014). A set of game players preceding to the player i for the kth resource at order θ is

$$\mathfrak{A}_i^\theta(k) = \{j \in \mathcal{B} : \theta(j) < \theta(i)\}.$$

Therefore, $v(\mathfrak{A}_i^\theta(k))$ can be expressed as

$$v(\mathfrak{A}_i^\theta(k)) = U\_\mathcal{P}_i^k \times \left[\sum_{q \in \mathfrak{A}_q^{\theta'}(k)} \partial_q^k - \sum_{q \in \mathfrak{A}_q^{\theta'}(k)} \mathcal{A}_q^k\right]^+ - U\_\mathcal{P}_i^k \times \left[\sum_{q \in \mathfrak{A}_q^{\theta'}(k)} \mathcal{A}_q^k - \sum_{q \in \mathfrak{A}_q^{\theta}(k)} \partial_q^k\right]^+$$

s.t.,

$$[x]^+ = \max(x, 0) \tag{12}$$

A marginal contribution of the player i at order θ in $(\mathcal{B}, v(\cdot), k)$ is defined by

$$\mathcal{S}_i^\theta(\mathcal{B}, v, k) = v(\mathfrak{A}_i^\theta(k) \cup \{i\}) - v(\mathfrak{A}_i^\theta(k)).$$

Then the *SV* of $(\mathcal{B}, v(\cdot), k)$ is defined as follows (Kamijo, 2009):

$$SV_i(\mathcal{B}, v, k) = \frac{1}{|\Theta(\mathcal{B})|} \times \sum_{\theta \in \Theta(\mathcal{B})} \left(\mathcal{S}_i^{\theta}(\mathcal{B}, v, k)\right), \text{ for all } i \in \mathcal{B} \quad (13)$$

where $|\cdot|$ represents the cardinality of the set. Therefore, the *SV* is an average of marginal contribution vectors where each order $\theta \in \Theta(\mathcal{B})$ occurs in an equal probability, that is, $1 / |\Theta(\mathcal{B})|$. Under the cooperative game situation, *SV* provide a unique solution with the desirable properties,

1. Efficiency,
2. Symmetry,
3. Additivity,
4. Dummy (Kamijo, 2009).

Although the *SV* is quite an interesting concept, and provides an optimal and fair solution for many applications, its main drawback is its computational complexity: the number of computations will increase prohibitively when the number of game players increases. Therefore, applications that utilize the *SV* remain scarce. In the HGRS scheme, if all possible orderings of *SBSs* ($\Theta(\mathcal{B})$) have to be taken into account in calculating equation (12) and (13), the computational complexity of calculating the *SV* can be very high and too heavy to be implemented in real *C-RAN* operations. To resolve this problem, the HGRS scheme adopts the new concept of *Asymptotic Shapley Value* (*A_SV*) approach, which is an approximation method for the *SV* under a large number of players (Kamijo, 2009; Lee, 2014). For the kth resource, let the *A_SV* of player i be ϕ_i^k; it is given as.

$$\phi_i^k = \begin{cases} \left(U\_\mathcal{P}_i^k \times \int_0^1 erf\left(\frac{\sqrt{\mathcal{P}_u^k} \times \tau}{\sqrt{2} \times \eta}\right) dp\right) \times \left(\partial_i^k - \mathcal{A}_i^k\right), & if\ \frac{\mu_s^k \times N_s^k}{\mu_B^k \times N_B^k} = 1 \\ U\_\mathcal{P}_i^k \times \left(\partial_i^k - \mathcal{A}_i^k\right), & if\ \frac{\mu_s^k \times N_s^k}{\mu_B^k \times N_B^k} \neq 1 \end{cases}$$

s.t.,

$$erf\left(x\right)=\frac{1}{\sqrt{\left|\Theta\left(\mathcal{B}\right)\right|}}\int_{-x}^{x}e^{-y^2}dy,\eta=\sqrt{\frac{\mu_B^k\times\left(\sigma_S^k\right)^2+\mu_s^k\times\left(\sigma_B^k\right)^2}{\mu_B^k+\mu_s^k}},\tau=\frac{\mu_s^k\times N_s^k-\mu_B^k\times N_B^k}{\sqrt{\mathcal{B}}} \quad (14)$$

where N_s^k and N_B^k are the number of players with the condition of $\partial^k - \mathcal{A}^k < 0$, and the condition of $\partial^k - \mathcal{A}^k \geq 0$, respectively. μ_s^k and μ_B^k (*or*$\left(\sigma_S^k\right)^2$ and $\left(\sigma_B^k\right)^2$) are the mean (*or* variance) of total wasted and needed k^{th} resource, respectively. The method for obtaining the proof of the derivation of *A_SV* value can be found in (Lee, 2014).

Under dynamic *C-RAN* environments, fixed resource distribution methods cannot effectively adapt to changing system conditions. The HGRS scheme treats the resource distribution for multiple *SBSs* as an on-line decision problem. At the time period t, the total amount of available k^{th} resource ($\mathcal{A}\_\mathcal{R}_t^k$) is dynamically re-distributed over *SBSs* according to ϕ^k values. In order to apply the time-driven implementation of resource re-distribution, the HGRS scheme partitions the time-axis into equal intervals of length *unit_time*. At the end of each time period, the re-distributed k^{th} resource amount for the SBS_i ($\Pi_i^k(t)$) is obtained periodically as follows.

$$\Pi_i^k\left(t\right)=\mathcal{A}\_\mathcal{R}_t^k\times\frac{\phi_i^k+\left|\min_{j\in\mathcal{B}}\left\{\phi_j^k\right\}\right|}{\sum_{b\in\mathcal{B}}\left(\phi_b^k+\left|\min_{j\in\mathcal{B}}\left\{\phi_j^k\right\}\right|\right)},s.t.,t\in T \quad (15)$$

C-RAN Resource Sharing in Lower Indian Buffet Game

In the lower-level Indian game model, multiple *MUs* requests different resources to their corresponding *SBS*. Let MU_i^j are the *MU* j in the area of SBS_i and $\mathcal{L}_i$ denote a set of resources in the ith*SBS*, $\mathcal{L}_i$ = {$\mathcal{R}_i^1$ = CPU, $\mathcal{R}_i^2$ = memory, $\mathcal{R}_i^3$ = storage, $\mathcal{R}_i^4$ = bandwidth}. Each $\mathcal{R}_i^k$ represents the available amount of kth resource in the SBS_i; these resources are obtained from the *VBP* through the upper-level Indian game. Individual *MU* attempts to actually purchase multiple resources based on their unit prices $U\_\mathcal{P}_i^k$, where $1 \le k \le 4$ and $i \in \boldsymbol{\mathcal{B}}$.

The lower-level Indian game deals with the resource allocation problem while maximizing resource efficiencies. Based on the reciprocal relationship between *SBSs* and *MUs*, the HGRS scheme adaptively allocates *SBSs*' resources to each *MU*. From the viewpoint of *MUs*, their payoffs correspond to the received benefit minus the incurred cost (Wang, 2005). Based on its expected payoff, each *MU* attempts to find the best actions. The *MU* j's utility function of kth resource (U_j^k) in the ith*SBS* is defined as.

$$U_j^k\left(\xi_j^k\left(i\right)\right) = b_j\left(\xi_j^k\left(i\right)\right) - c\left(U\_\mathcal{P}_i^k, \xi_j^k\left(i\right)\right)$$

s.t.,

$$b_j\left(\xi_j^k\left(i\right)\right) = \omega_j^k \times \log\left(\xi_j^k\left(i\right)\right) \text{ and } mp^k \le U\_\mathcal{P}_i^k \le Mp^k \tag{16}$$

where $\xi_j^k\left(i\right)$ is the *MU* j's requested amount of kth resource in the SBS_i, and $b_j\left(\xi_j^k\left(i\right)\right)$ is the received benefit for the *MU* j. ω_j^k represents a payment that the *MU* j would spend for the kth resource based on its perceived worth. The $U\_\mathcal{P}_i^k$ is the unit price for the kth resource unit in the SBS_i, and $c\left(U\_\mathcal{P}_i^k\left(i\right), \xi_j^k\left(i\right)\right)$ is the cost function of SBS_i. Each *SBS* decides the $U\_\mathcal{P}_i^k$ between the pre-defined minimum (mp^k) and the maximum (Mp^k) price boundaries. In general, a received benefit typically follows a model of diminishing returns to scale; *MU*'s marginal benefit diminishes with increasing bandwidth (Wang, 2005). Based on this consideration, the received benefit can be represented in a general form of *log* function. In a distributed self-regarding fashion, each individual *MU* is independently interested in the sole goal of maximizing his/her utility function as.

$$\max_{\xi_j^k(i) \ge 0} U_j^k\left(\xi_j^k\left(i\right)\right) = \max_{\xi_j^k(i) \ge 0} \left\{b_j\left(\xi_j^k\left(i\right)\right) - c\left(U\_\mathcal{P}_i^k\left(i\right), \xi_j^k\left(i\right)\right)\right\} \tag{17}$$

From the viewpoint of *SBSs*, the most important criterion is a total revenue; it is defined as the sum of payments from *MUs* (Feng, 2004). Based on the $U\_\mathcal{P}_i^k$ and the total allocated resource amounts for *MU*'s, the total revenue of all *SBSs* (Ψ) is given by

$$\Psi = \sum_{i=1}^{n} \Psi_i = \sum_{i=1}^{n}\sum_{k=1}^{l}\left(U\_\mathcal{P}_i^k \times T_i^k\right) = \sum_{i=1}^{n}\sum_{k=1}^{l}\sum_{j=1}^{m}\left(U\_\mathcal{P}_i^k \times \xi_j^k(i) \times \iota_j^k(i)\right)$$

s.t.,

$$\iota_j^k(i) = \begin{cases} 1, & \textit{if the requested } \xi_j^k(i) \textit{ is actually allocated} \\ 0, & \textit{otherwise} \end{cases} \quad (18)$$

where n, l, m are the total number of *SBSs*, resources and *MUs*, respectively. Each *SBS* adaptively controls its own $U\_\mathcal{P}^k$ to maximize the revenue in a distributed manner. The traffic model is assumed based on the elastic-demand paradigm; according to the current $U\_\mathcal{P}^k$, *MUs* can adapt their resource requests. It is relevant in real world situations where *MUs*' requests may be influenced by the price (Yang, 1997a; Yang, 1997b). In response to ω_j^k, the *MU* j can derive the $\xi_j^k(i)\left(= \omega_j^k / U\_\mathcal{P}_i^k\right)$. In the SBS_i, the total requested resource amount from corresponding *MUs* is defined as.

$$\sum_{k=1}^{l}\sum_{j=1}^{m}\xi_j^k(i) = \sum_{k=1}^{l}\left(\frac{\sum_{j=1}^{m}\omega_j^k}{U\_\mathcal{P}_i^k}\right) \quad (19)$$

When the price is low, more *MUs* are attracted to participate *C-RAN* services because of the good satisfactory payoff. However, if the price is high, *MUs* requests are reduced because of the unsatisfactory payoff. Therefore, to deal with the congestion problem, a higher price is suitable to match the resource capacity constraint while reducing the potential demands. In order to toward the demand-supply balance, the current price should increase or decrease by $\Delta U\_\mathcal{P}^k$.

In the HGRS scheme, *SBSs* individually take account of previous price strategies to update their beliefs about what is the best-response price strategy in the future. If a strategy change can bring a higher payoff, *SBSs* have a tendency to move in the direction of that successful change, and vice versa. Therefore, *SBSs* dynamically tune their current strategies based on the payoff history. For the kth resource, the SBS_i's price strategy at the time period $t+1$ ($\lambda_i^k(t+1)$) is defined as

$$\begin{cases} \lambda_i^k(t+1) = \Lambda\left[\lambda_i^k(t) + \left|\Delta U\_\mathcal{P}_i^k(t)\right|\right], & If\ \Omega > 0 \\ \lambda_i^k(t+1) = \Lambda\left[\lambda_i^k(t) - \left|\Delta U\_\mathcal{P}_i^k(t)\right|\right], & If\ \Omega \leq 0 \end{cases}$$

s.t.,

$$\Delta U\_\mathcal{P}_i^k(t) = \frac{\left(\Psi_i^k(t) - \Psi_i^k(t-1)\right)}{\Psi_i^k(t-1)},\ \Omega = \frac{\left(\lambda_i^k(t) - \lambda_i^k(t-1)\right)}{\Delta U\_\mathcal{P}_i^k(t)}$$

and

$$\mathbf{\Lambda}\left[\mathcal{K}\right] = \begin{cases} \Lambda\left[\mathcal{K}\right] = mp^k, & if\ \mathcal{K} < mp^k \\ \Lambda\left[\mathcal{K}\right] = \mathcal{K}, & if\ mp^k \leq \mathcal{K} \leq Mp^k \\ \Lambda\left[\mathcal{K}\right] = Mp^k, & if\ \mathcal{K} > Mp^k \end{cases} \tag{20}$$

where $\Psi_i^k(t)$ and $\lambda_i^k(t)$ are the SBS_i's revenue and price strategy for the k th resource at the time period t, respectively. $\left|” U\_\mathcal{P}_i^k(t)\right|$ represents the absolute value of $” U\_\mathcal{P}_i^k(t)$. According to (20), the strategy profile of all *SBSs* can be denoted by a $n \times l$ matrix as follows:

$$\mathcal{T}_i(t) = \left\{\lambda_i^1(t), \lambda_i^2(t), \ldots. \lambda_i^l(t)\right\} = \begin{bmatrix} \lambda_1^1(t), \lambda_1^2(t), \cdots, \lambda_1^l(t) \\ \lambda_2^1(t), \lambda_2^2(t), \cdots, \lambda_2^l(t) \\ \vdots \quad\quad \vdots \quad\quad \ddots \quad \vdots \\ \lambda_n^1(t), \lambda_n^2(t), \cdots, \lambda_n^l(t) \end{bmatrix}., \text{ s.t., } i \in \boldsymbol{\mathcal{B}} \tag{21}$$

The Main Steps of the HGRS Scheme

The HGRS scheme presents the two-level Indian buffet game model. In the upper-level Indian buffet game, available resources of *CPs* are distributed to *SBSs* based on the concept of *A_SV*. In the lower-level Indian buffet game, Individual *SBS* allocate the distributed resources to *MUs* according to the non-cooperative manner. The main steps of the HGRS scheme are given next.

Step 1: At the initial time, all *SBSs* have same price strategies ($\mathcal{T}$). At the beginning of game, this starting guess is a reasonable assumption.

Step 2: At each game period, the *VBP* collects available resources from *CPs* using the virtualization technology, and distribute these resources to each *SBS* according to (12)-(15).

Step 3: Individual *MU* in each cell attempts to actually purchase multiple resources from corresponding *SBS*. Based on this information, each *SBS* dynamically decide the price strategy ($\mathcal{T}$) using (19) and (20).

Step 4: At each game period, the *VBP* re-distributes periodically the *CP* resources based on the currently calculating ϕ values; it is the upper-level Indian game.

Step 5: Based on the current price ($\mathcal{T}$), each *MUs* dynamically decide the amount of purchasing resources according to (17).

Step 6: Strategy decisions in each game player are made in an entirely distributed manner.

Step 7: Under widely diverse *C-RAN* environments, the *VBP*, *SBSs* and *MUs* are self-monitoring constantly based on the iterative feedback mechanism.

Step 8: If the change of prices in all *SBSs* is within a pre-defined bound (ε), this change is negligible; proceed to Step 9. Otherwise, proceed to Step 2 for the next iteration.

Step 9: Game is temporarily over. Ultimately, the HGRS scheme reaches an effective resource sharing solution. When the *C-RAN* system status is changed, it can re-trigger another game-based resource sharing procedure.

Summary

Efficient and fine-grained resource sharing becomes an increasingly important and attractive control issue for new-generation *C-RAN* systems. The HGRS scheme proposes a novel multi-resource sharing algorithm, which is framed as a two-level Indian buffet game model: upper-level Indian game is played among *VBP-SBSs*, and lower-level Indian game is played among *SBSs-MUs*. Based on the hierarchical interaction mechanism, the *VBP*, *SBSs* and *MUs* are intertwined and make decisions during the step-by-step interactive feedback process. The main novelty of HGRS scheme is to apply a new resource sharing paradigm to control the *C-RAN* environment.

CLOUD-BASED COMPUTATION OFFLOADING (CCO) SCHEME

The *Cloud-based Computation Offloading* (CCO) scheme is designed as an effective MCIoT computation offloading algorithm for Future IoT Platform. Based on the nested game model, each mobile device determines the portion of remote offloading computation based on the Rubinstein game approach. Then, a computation resource in the cloud system is dynamically assigned for the requested offloading computation. The CCO scheme can approach an optimal solution for the offloading computation in MCIoT system.

Development Motivation

Recently, the rapid technology development makes it possible for connecting various smart mobile devices together while providing more data interoperability methods for application purpose. Therefore, the diverse nature of applications has challenged the communication and computation mechanisms to look beyond conventional applications for effective network policies, Quality of Service (QoS) and system performance. The Internet of Things (IoT) paradigm is based on intelligent and self configuring mobile devices interconnected in a dynamic and global network infrastructure. It is enabling ubiquitous and pervasive computing scenarios in a real world (Botta, 2014; Kim, 2010).

In the current decade, a growing number of researches have been conducted to acquire data ubiquitously, process data timely, and distribute data wirelessly in the IoT paradigm. To satisfy this requirement, mobile devices should have a capacity to handle the required processing and computation work. Unfortunately, the desire for rich, powerful applications on mobile devices conflicts with the reality of these devices limitations: slow computation processors, little memory storage and limited battery life. For this reason, mobile devices still lag behind desktop and server hardware to provide the experience that users expect. (Singh, 2014; Vermesan, 2011).

The Mobile Cloud (MC) is emerging as one of the most important branches of cloud computing and is expected to expand the mobile ecosystems. Usually, cloud computing has long been recognized as a paradigm for big data storage and analytics; it has virtually unlimited capabilities in terms of storage and processing power. MC is the combination of cloud computing, mobile computing and wireless networks to bring rich computational resources to mobile users, network operators, as well as cloud computing providers. With an explosive growth of the multimedia mobile applications, MC computing has become a significant research topic of the scientific and industrial communities (Chen, 2015; Sabyasachi, 2013; Zhu, 2014).

The two fields of MC and IoT have been widely popular as future infrastructures and have seen an independent evolution. However, MC and IoT are complementary technologies, and several mutual advantages deriving from their integration have been identified (Singh, 2014). Therefore, a symbiosis has developed between mobile devices and MC, and is expected to be combined for the future Internet. Generally, mobile devices can benefit from the virtually unlimited capabilities and resources of MC to compensate its storage, processing, energy constraints. Specifically, the MC can offer an effective solution to implement IoT service management. On the other hand, the MC can benefit from the IoT system by extending its scope to deal with real world things in a more distributed and dynamic manner (Botta, 2014). Nowadays, the extension of MC over dynamic IoT environments has been referred as a next generation communication and computing paradigm (Sabyasachi, 2013).

The CCO scheme focuses on the integration of MC and IoT, which is call the MCIoT paradigm. MCIoT should support wide variety of multimedia applications with different QoS requirements; these applications need different system resources. However, mobile devices are generally characterized with limited storage and processing capacity. A possible way to dealing with this problem is to remotely execute some computation tasks on a more powerful cloud system, with results communicated back to the mobile devices. This method is the computation offloading (Sinha, 2011). Therefore, the synergy of MC and IoT lies at the junction of mobile devices, different wireless network providers and cloud computing systems.

Although the computation offloading approach can significantly augment computation capability of mobile devices, the task of developing a comprehensive and reliable computation offloading mechanism remains challenging. A key challenge is how to efficiently coordinate multiple mobile devices and MCIoT system. Usually, individual mobile devices locally make control decisions to maximize their profits in a distributed manner. This situation leads us into game theory. In 1988, an American political scientist, George Tsebelis introduced an important new concept, called nested games, to rational choice theory and to the study of comparative politics (Tsebelis, 1988). Using the notion of nested games, he showed that game players are involved simultaneously in several games. He argued that the '*nestedness*' of the principal game explains why a player confronted with a series of choices might not pick the alternative which appears to be optimal. In other words, what seems to be irrational in one arena becomes intelligible when the whole network of games is examined. Originally, the nested game has been used by anyone interested in the effects of political context and institutions on the behavior of political actors. Nowadays, nested game approach can be used to analyze a systematic, empirically accurate, and theoretically coherent account of apparently irrational actions (Jesse, 2002; Pang, 2011; Wang, 2013).

The CCO scheme adopts a nested game approach to address the computation offloading algorithm in the MCIoT platform. Nested game model is a useful framework for designing decentralized mechanisms, such that the mobile devices can self-organize into the mutually satisfactory computation offloading decisions. Usually, different mobile devices may pursue different interests, and act individually to maximize their profits. This self-organizing feature can add autonomics into MCIoT systems and help to ease the heavy burden of complex centralized control algorithms.

Nested Game Model for Computation Offloading

Under competitive or cooperative environments, behaviors of game players have direct influence on each other. Based on rational assumptions, game theory is to study the decision-making mechanism and the balance of decision-making interactions (Kim, 2014). However, game players seem to act irrationally, but if treating the game as a part of a larger game, we can see their behaviors are rational (Pang, 2011). From the view of small independent game, each player's strategy is not the optimal solution. However, from the view of big game, players' reactions are the best responses. Such games are called as nested games, and small games may be used as sub-games nested in the sequential game of a larger game (Jesse, 2002; Pang, 2011; Wang, 2013). In multiple fields of nested game, game players try to optimize their payoffs in the principal game field and also involved a game about the rules of the game. It can lead to apparently suboptimal payoffs as the game player fails to see the other fields that provide context for the small game in the principal field. Several studies use nested game theory to explain political party behavior, budget negotiations, electoral systems and public policy (Park, 2007).

The CCO scheme specifies the nested game to design a new computation offloading algorithm. The key point of MC offloading mechanism hinges on the ability to achieve enough computing resources with small energy consumption. In recent years, this technique has received more attention because of the significant rise of offload-available mobile applications, the availability of powerful clouds and the improved connectivity options for mobile devices. The main challenges to design an effective computation offloading algorithm lies in the adaptive division of applications for partial offloading, the mismatch control mechanism between how individual mobile devices demand and access computing resources, and how cloud providers offer them. To decide what, when and how to be offloaded, the CCO scheme should consider the offload overhead and current MCIoT system conditions.

In the resource-rich MC environment, a mobile device must pay the price to take advantage of computation offloading. To gain an extra benefit, idle computation resources in MCIoT system compete to get the requested offloading task. Therefore, mobile devices can select the most adaptable computation resource to execute their offloaded computations. The MCIoT environment can be described as follows:

1. $\mathbb{D} = \{D_0, D_1, \ldots, D_n\}$ is the set of mobile devices, and $A_i, i \in [1, n]$ is an application, which belongs to the mobile device D_i.
2. Mobile device applications are elastic applications, and can be split. For example, $A_i = \sum_{k=1}^{L} a_k^i$, where a_k^i is the *k*th module of A_i, and some parts (i.e., coded modules) of A_i can be offloaded.
3. $\mathbb{R} = \{R_0, R_1, \ldots, R_m\}$ is the set of idle cloud computing resources in the MCIoT environment, and $R_j, j \in [1, m]$ has a computation capacity ($\mathcal{F}_{R_j}^{R}$; CPU cycles per second) and expected price (ψ^{R_j}; price per CPU cycle of the R_j) to accomplish the offloaded computations.
4. Price in each $R_{j, 1 \le j \le m}$ can be dynamically adjustable according to the auction mechanism.
5. For simplicity, there is no communication noise and uncertainties in MCIoT environments.

The CCO scheme develops a two-stage nested game model comprised of elastic applications, mobile devices and computation resources in the MC. In the first stage, applications of mobile devices are divided into two parts: one part runs locally and the other part is run on the MC side. In the second stage, offloaded tasks are matched to computing resources in the MCIoT system. Based on the auction mechanism, computation resources submit different offers to get the requested offload task, and the most adaptable offer is selected. According to the two-stage sequential nested game approach, the CCO scheme can make decisions about whether to perform computation offloading, which portion of application should be offloaded to the cloud, and which resource is selected to accomplish the requested offload.

Offloading Communication and Computation Process

By taking into account both communication and computation aspects of MCIoT environments, the CCO scheme formulate a new decentralized computation offloading algorithm. In each mobile device, applications can be computed either locally on the mobile device or remotely on the cloud via computation offloading. For the local computing approach, each mobile device (e.g., D_i) can execute the some computation part of A_i, individually. The local computation execution time ($L\_CT_{comp}^{A_i-D_i}$)) of the application A_i on the mobile device D_i is given as

$$L\_CT_{comp}^{A_i-D_i} = \frac{\sum_{k=1}^{L} \mathfrak{U}\left(a_k^i\right) \times a_k^i}{\mathcal{F}_{D_i}^{L}}, s.t., \mathfrak{U}\left(a_k^i\right) = \begin{cases} 1, & if\ a_k^i\ is\ locally\ computed \\ 0, & otherwise\left(a_k^i\ is\ offloaded\right) \end{cases} \quad (22)$$

where $\mathcal{F}_{D_i}^{L}$ is the computation capability of mobile device D_i. The local computation cost ($L\_CC_{comp}^{A_i-D_i}$) of the application A_i on the mobile device D_i is calculated based on the $L\_CT_{comp}^{A_i-D_i}$ and consumed local computation energy (ρ).

$$L\_CC_{comp}^{A_i-D_i} = \rho^{D_i} \times \left(\mathcal{F}_{D_i}^{L} \times L\_CT_{comp}^{A_i-D_i}\right) \quad (23)$$

where ρ^{D_i} is the coefficient denoting the consumed energy cost per CPU cycle. The evaluation of total local computation overhead ($T\_O_{local}^{A_i-D_i}$) of the application A_i on the mobile device D_i is a non-trivial multi-objective optimization problem. It is addressed as a weighted sum by considering normalized time and energy cost.

$$T\_O_{local}^{A_i-D_i} = \lambda_{D_i}^{A_i} \times \left(\frac{L\_CT_{comp}^{A_i-D_i}}{\frac{1}{\mathcal{F}_{D_i}^{L}} \times \sum_{k=1}^{L} a_k^i} \right) + \left(1 - \lambda_{D_i}^{A_i}\right) \times \left(\frac{L\_CC_{comp}^{A_i-D_i}}{\rho^{D_i} \times \sum_{k=1}^{L} a_k^i} \right) \quad (24)$$

where $\lambda_{D_i}^{A_i}$ is a parameter to control the relative weights given to execution time and energy consumption. To satisfy the A_i's demand, $\lambda_{D_i}^{A_i}$ is adaptively decided. In the equation (30), the $\lambda_{D_i}^{A_i}$ value decision process is explained in detail.

Next, the CCO scheme estimates the remote computation overhead through the MC offloading mechanism. Generally, the communication and computation aspects play a key role in MC offload. The CCO scheme considers a delay sensitive Wi-Fi model for offloading services; mobile devices are sensitive to delay and their payoff decreases as delay increases. As a wireless access Base-Station (BS), a WiFi access point manages the uplink/downlink communications of mobile devices. For the computation offloading, the mobile device D_i would incur the extra overhead in terms of time and energy to submit the computation offload via wireless access. Based on the communication model in (Chen, 2015), the offloading communication time ($O\_CT_{off}^{A_i-D_i}$) and energy ($O\_CE_{off}^{A_i-D_i}$) of the application A_i on the mobile device D_i are computed as follows.

$$O\_CT_{off}^{A_i-D_i}(\mathbb{D}) = \frac{\sum_{k=1}^{L} \mathfrak{T}(a_k^i) \times a_k^i}{\mathcal{B} \times \log_2\left(1 + \frac{P_i \times H_{i,BS}}{\omega + \sum_{a_k^g \in A_g \setminus \{A_i\}: \mathfrak{T}(a_k^g)=1} P_g \times H_{g,BS}}\right)}$$

$$s.t., g \in [1,n], D_g \in \mathbb{D} \text{ and } \mathfrak{T}(a_k^i) = \begin{cases} 1, & if\ a_k^i\ is\ offloaded \\ 0, & otherwise \end{cases} \quad (25)$$

$$O\_CE_{off}^{A_i-D_i}(\mathbb{D}) = \frac{P_i \times \sum_{k=1}^{L} \mathfrak{T}(a_k^i) \times a_k^i}{\mathcal{B} \times \log_2\left(1 + \frac{P_i \times H_{i,BS}}{\omega + \sum_{a_k^g \in A_g \setminus \{A_i\}: \mathfrak{T}(a_k^g)=1} P_g \times H_{g,BS}}\right)}$$

where $\mathcal{B}$ is the channel bandwidth and P_i is the transmission power of device D_i. $H_{i,BS}$ denotes the channel gain between the mobile device D_i and the BS, and ω denotes the interference power. From the equation (25), the CCO scheme can see that if too many mobile devices choose to offload the computation via wireless access simultaneously, they may incur severe interference, leading to low data rates. It would negatively affect the performance of MC communication. Therefore, offloading decisions among mobile devices are tightly coupled each other (Chen, 2015). To address this conflicting situation, game theory can be adopted to achieve efficient computation offloading decisions.

After the offloading, the computation time ($C\_T_{remote}^{A_i\_R_j}$) and payment ($P_{remote}^{A_i\_R_j}$) of remote computation task ($\sum_{k=1}^{L}\mathfrak{T}\left(a_k^i\right)\times a_k^i$) on the assigned computation resource R_j can be then given as

$$C\_T_{remote}^{A_i\_R_j} = \frac{\sum_{k=1}^{L}\mathfrak{T}\left(a_k^i\right)\times a_k^i}{\mathcal{F}_{R_j}^{R}} \text{ and } P_{remote}^{A_i\_R_j} = \psi^{R_j}\times\mathcal{F}_{R_j}^{R}\times C\_T_{remote}^{A_i\_R_j}. \tag{26}$$

where $\mathcal{F}_{R_j}^{R}$ is the R_j's computation capability and ψ^{R_j} is the coefficient denoting the price per CPU cycle of the R_j. According to (25) and (26), the total offload overhead ($T\_O_{off}^{A_i\_R_j}$) of the application A_i on the computation resource R_j is computed as a weighted sum by considering execution time and consuming cost.

$$T\_O_{off}^{A_i\_R_j} = \lambda_{D_i}^{A_i}\times\left(\frac{O\_CT_{off}^{A_i\_D_i}\left(\mathbb{D}\right)+C\_T_{remote}^{A_i\_R_j}}{\frac{1}{\mathcal{F}_{D_i}^{L}}\times\sum_{k=1}^{L}a_k^i}\right)+\left(1-\lambda_{D_i}^{A_i}\right)\times\left(\frac{O\_CE_{off}^{A_i\_D_i}\left(\mathbb{D}\right)+P_{remote}^{A_i\_R_j}}{\rho^{D_i}\times\sum_{k=1}^{L}a_k^i}\right) \tag{27}$$

According to (24), (25), (26) and (27), the total execution time of A_i ($T\_C_{total}^{A_i\_D_i,R_j}$) with partial offloading can be estimated considering between the local and remote computing times.

$$T\_C_{total}^{A_i\_D_i,R_j} = \max\left[L\_CT_{comp}^{A_i\_D_i},\left(O\_CT_{off}^{A_i\_D_i}\left(\mathbb{D}\right)+C\_T_{remote}^{A_i\_R_j}\right)\right] \tag{28}$$

Finally, we can compute the total execution overhead of A_i ($\mathcal{S}^{A_i\_D_i,R_j}\left(\mathbb{D}\right)$) like as

$$\mathcal{S}^{A_i\_D_i,R_j}(\mathbb{D}) = \lambda_{D_i}^{A_i} \times \left(T\_C_{total}^{A_i\_D_i,R_j}\right) + \left(1-\lambda_{D_i}^{A_i}\right) \times \left(\frac{L\_CC_{comp}^{A_i\_D_i} + \left(O\_CE_{off}^{A_i\_D_i}(\mathbb{D}) + P_{remote}^{A_i\_R_j}\right)}{\rho^{D_i} \times \sum_{k=1}^{L} a_k^i}\right) \quad (29)$$

To meet the application-specific demand, different applications have different evaluation criteria for time and energy consumption. For example, when a mobile device is running an application that is delay sensitive (e.g., real-time applications), it should put more weight on the execution time (i.e., a higher λ) in order to ensure the time deadline, and vice versa. Therefore, a fixed value for λ cannot effectively adapt to the different application demands. In this work, the value of λ for the A_i on mobile device D_i ($\lambda_{D_i}^{A_i}$) is dynamically decided as follows.

$$\lambda_{D_i}^{A_i} = \min\left[1, \left(\frac{\frac{1}{\mathcal{F}_{D_i}^{L}} \times \sum_{k=1}^{L} a_k^i}{T\_D^{A_i}}\right)\right] \quad (30)$$

where $T\_D^{A_i}$ is the time deadline of A_i. Therefore, through the real-time online monitoring, the CCO scheme can be more responsive to application demands.

Application Partitioning Game

An important challenge for partial offloading is how to partition elastic applications and which part of partitioned application should be pushed to the remote clouds. The CCO scheme analyzes how mobile devices can exploit a partial offloading between cloud computation and local computation. To distribute computation tasks for partial offloading, the CCO scheme provides a non-cooperative bargaining game model by considering the consuming cost and computation time. Usually, a solution to the bargaining game model enables the game players to fairly and optimally determine their payoffs to make joint-agreements (Park, 2007; Suris, 2007). Therefore, the bargaining model is attractive for the partitioning problem.

In *Rubinstein-Stahl* model, players have their own bargaining power (δ). They negotiate with each other by proposing offers alternately. After several rounds of negotiation, players finally reach an agreement as following (Pan, 2008; Zhao, 2002).

$$\left(x_1^*, x_2^*\right) = \begin{cases} \left(\dfrac{1-\delta_2}{1-\left(\delta_1 \times \delta_2\right)}, \dfrac{\delta_2 \times \left(1-\delta_1\right)}{1-\left(\delta_1 \times \delta_2\right)}\right) & \textit{if the player\_1 offers first} \\ \left(\dfrac{\delta_1 \times \left(1-\delta_2\right)}{1-\left(\delta_1 \times \delta_2\right)}, \dfrac{1-\delta_1}{1-\left(\delta_1 \times \delta_2\right)}\right) & \textit{if the player\_2 offers first} \end{cases}$$

s.t.,

$$\left(x_1^*, x_2^*\right) \in \boldsymbol{R}^2 : x_1^* + x_2^* = 1, x_1^* \geq 0, x_2^* \geq 0 \text{ and } 0 \leq \delta_1, \delta_s \leq 1 \tag{31}$$

It is obvious that

$$\frac{1-\delta_2}{1-\left(\delta_1 \times \delta_2\right)} \geq \frac{\delta_2 \times \left(1-\delta_1\right)}{1-\left(\delta_1 \times \delta_2\right)}$$

and

$$\frac{\delta_1 \times \left(1-\delta_2\right)}{1-\left(\delta_1 \times \delta_2\right)} \leq \frac{1-\delta_1}{1-\left(\delta_1 \times \delta_2\right)}.$$

Traditionally, the bargaining power in the *Rubinstein-Stahl's* model is defined as follows (Pan, 2008).

$$\delta = e^{-\xi \times \Phi}, \text{ s.t., } \xi > 0 \tag{32}$$

where Φ is the time period of negotiation round. Given the Φ is fixed, δ is monotonic decreasing with ξ. Therefore, ξ is an instantaneous discount factor to adaptively adjust the bargaining power.

In the CCO scheme, the *Rubinstein-Stahl* bargaining game model is formulated to solve the application partitioning problem. In the game model, cloud computation resource and mobile device are assumed as players, which are denoted as the *player_1* (i.e., mobile device for local computation) and *player_2* (i.e., cloud resource for remote computation). In the scenario of the *Rubinstein-Stahl* model, each player has different discount factor (ξ). Under various MCIoT situations, the CCO scheme dynamically adjusts ξ values to provide more efficient control over system condition fluctuations. When the current local computation overhead is heavy, the mobile device does not have sufficient computation capacity to support the local computation service. In this case, a higher value of the *player_1*'s discount factor (ξ_I) is more suitable. If the reverse has been the case (i.e., the remote computation overhead is heavy), a higher value of the *player_2*'s discount factor (ξ_{II}) is suitable. At the end of each game period, the *player_1* and *player_2* adjust their discount factor values (ξ_I and ξ_{II}) as follows.

$$\xi_I = 1 - \xi_{II}, s.t., \xi_{II} = \frac{T\_O_{off}^{A_i - R_j}}{\mathcal{S}^{A_i\_D_i, R_j}(\mathbb{D})} \tag{33}$$

For simplicity, the CCO scheme assumes that the ξ values are fixed within an offloading procedure for each application, while they can be changed in different applications. Therefore, as system situations change after application partitioning, each player can adaptively adjust their ξ_I and ξ_{II} values for the next application execution while responding current MCIoT system conditions.

Cloud Resource Selection Game

Recently, researchers have proposed various auction models to optimally match up the buyer and seller according to their desires. It is a significant and efficient market-based approach to solve the allocation problem with more requisitions. Therefore, the auction game model can provide a resource selection mechanism in MC systems. In the computation resource selection scenario, there are a requested offload task (i.e., buyer) and computation resources (i.e., sellers). Based on the sequential offloading requests, the action model is designed as the one-to-many auction structure. As sellers, computation resources ($\mathbb{R}$) in MC system offer bids (i.e., the expected selling prices) for the remote offload computation. To show their preference to get the offloading computation, sellers ($\mathbb{R}$) can adjust their selling prices, periodically. For the tth auction stage, the seller (i.e., $R_j \in \mathbb{R}$) bids his price ($\psi^{R_j}(t)$) per CPU cycle is defined as follows.

$$\psi^{R_j}(t) = \begin{cases} \psi^{R_j}(t-1) + \left(1 - \dfrac{1}{\left(\exp\left(\max\left[5-0, \varepsilon_{R_j}\right]\right)\right)}\right), & \textit{if } R_j \textit{ is selected at } t-1 \\ s.t., \varepsilon_{R_j} \sim N\left(\mu_{R_j}, \sigma^2_{R_j}\right) \\ \psi^{R_j}(t-1) - \left(1 - \dfrac{1}{\left(\exp\left(\max\left[0, \varepsilon_{R_j}\right]\right)\right)}\right), & \textit{otherwise} \end{cases}$$

(34)

where ε_{R_j} is a random variable to presents the price adjustment. Because the sellers are not interrelated, the random variable (ε) of each seller is independent of each other. According to (34), each seller (i.e., $R_j \in \mathbb{R}$) bids his offer (ψ^{R_j}, $\mathcal{F}^R_{R_j}$) at each auction round, and then the buyer (i.e., $D_i \in \mathbb{D}$) select the most adaptable offer. In the CCO scheme, the minimum price offering resource while satisfying the computation deadline ($T\_D^{(\cdot)}$) is selected. This dynamic auction procedure is repeated sequentially serial auction round. In each auction round, sellers can learn the buyer's desire with incoming information, and can make a better price decision for the next auction.

The Main Steps of the CCO Scheme

The CCO scheme presents a two-stage nested game model for the interaction of game players such as elastic applications, mobile devices and MC computation resources. Applications are involved in the application partitioning game in the first stage, and mobile devices and MC resources are involved in the resource selection game in the second stage. This two-stage nested game reflects the sequential dependencies of decisions in each stage. The main steps of the CCO scheme are given next.

Step 1: At the initial time, applications in each mobile devices are equally partitioned for local and remote offloading computations. At the beginning of game, this starting guess is useful to monitor the current MCIoT situation.

Step 2: According to (22)-(30), mobile devices can estimate their total offload overhead ($T\_O_{off}^{(\cdot)}$) and the total execution overhead ($\mathcal{S}^{(\cdot)}(\mathbb{D})$), individually.

Step 3: Based on the (31)-(33) equations, each mobile device adaptively re-partitions its application based on the *Rubinstein-Stahl* bargaining game model.

Step 4: One part is computed locally on the mobile device. In MC side, the other part is computed remotely on the computation resource, which is selected according to the auction game.

Step 5: For the next resource selection process, computation resources periodically adjust their selling prices ($\psi^{(\cdot)}$) according to (34).

Step 6: As game players, elastic applications, mobile devices and MC computation resources are interrelated, and interact with each other in a two-stage nested game. In each stage game, game players try to maximize their payoffs while they are involving in a bigger game.

Step 7: Under widely diverse MCIoT environments, mobile devices and computation resources are self-monitoring constantly for the next iterative feedback processing. This iterative feedback procedure continues under MCIoT system dynamics.

Step 8: When a new application service is requested, it can re-trigger another computation offloading process; proceeds to Step 1 for the next game iteration.

Summary

Over recent past years, a novel paradigm where Cloud and IoT are merged is expected to be an important component of the Future Internet. The CCO scheme reviews the integration of MC and IoT, and designs a new computation offloading algorithm in the MCIoT platform. Based on the nested game model, the main goal of the CCO scheme is to maximize mobile device performance while providing service QoS. To satisfy this goal, the CCO scheme consists of application partitioning game and cloud resource selection game. In the partitioning game, applications are adaptively partitioned according to the *Rubinstein-Stahl* bargaining model. In the resource selection game, computation resources in MC system are selected based on the one-to-many auction game model. Based on the nested game principle, these game models are interrelated to each other, and operated as a two-stage sequential game.

DYNAMIC SOCIAL CLOUD MANAGEMENT (DSCM) SCHEME

The *Dynamic Social Cloud Management* (DSCM) scheme is developed with a view of game theory model and reciprocal resource sharing mechanism. In particular, the DSCM scheme devises a new transformable Stackelberg game to coordinate the interdependence between social structure and resource availability for individual users. The DSCM scheme constantly monitors the current SC system conditions and adaptively exploits the available resources while ensuring mutual fairness.

Development Motivation

Digital relationships between individual people become more and more embedded in our daily actions, and they can be powerful influences in our real-life. Moreover, we are now connected with all our social networks through mobile devices. The increasing ubiquity of social networks is evidenced by the growing popularity of social network services. A social network service consists of a representation of each user, his or her social links, and a variety of additional services. Usually, social networks provide a platform to facilitate communications and resource sharing between users while modelling real-world relationships. Therefore, a variety of social network services have extended beyond simple communication among users (Hemmati, 2010; Jang, 2013; Zhang, 2014).

With the advent of social networks, cloud computing is becoming an emerging paradigm to provide a flexible stack of computing, software, and storage services. In a scalable and virtualized manner over networks, cloud users can access to fully virtualized hardware resources. The adoption of cloud computing technology is attractive; users obtain cloud resources, whose management is partly automated and can be scaled almost instantaneously. However, with the rapid development of cloud computing, critical issues of cloud computing technology have emerged. In general, modern cloud applications are characterized by assuming a constant environment. But, real-world environments are open, dynamic and unpredictable (Chen, 2015; Liu, 2015).

In social networks, individual users are bound by finite resource capacity and limited capabilities. However, some users may have surplus resource capacity or capabilities. Therefore, the superfluous resource could be shared for a mutual benefit. Within the context of a social network, users may wish to share resources without payment, and utilize a reciprocal credit based on the trust model (Chard, 2010; Chard, 2012). To satisfy this goal, a new concept, Social Cloud (SC) was introduced by combining the methodologies of social networks and cloud computing. SC is a novel scalable computing model where resources are beneficially shared among a group of social network users. From (Chard, 2010), the formal definition of SC can be defined like as; *A social cloud is a resource and service sharing framework utilizing relationships established between members of a social network.* Based on the cloud computing technique, SC model is used to enable virtualized resource sharing through service-based interfaces.

To construct the SC system in a real-world environment, there are many challenges that need to be carefully considered. First of all, the concept of SC focuses on the sharing rather than sale of resources. Using sharing preferences, the social context of exchange is accentuated along with the social ties of individual users. However, social relationships are not simply edges in a graph. There are many different types of relationship; different users will associate different levels of trust to different relationship contexts, and have different reliability, trustworthiness and availability. Therefore, users may have very specific preferences with whom they interact. To design an effective SC control scheme, it is necessary to take into account the preferences and perceptions of users towards one another (Caton, 2014).

Under widely dynamic SC system conditions, end users can be assumed as intelligent rational decision-makers, and they select a best-response strategy to maximize their expected payoffs. This situation is well-suited for the game theory. In 1934, German economist H. V. Stackelberg proposed a hierarchical strategic game model based on two kinds of different decision makers. Under a hierarchical decision making structure, one or more players declare and announce their strategies before the other players choose their strategies. In game theory terms, the declaring players are called as leaders while the players who react to the leaders are called as followers.

Originally, Stackelberg game model was developed to explain the monopoly of industry. The leader is the incumbent monopoly of the industry and the follower is a new entrant; it can be the static bilevel optimization model (Kim, 2014). The DSCM scheme has further extended the classical Stackelberg model and developed a novel game mode, called Transformable Stackelberg (TS) game model. In the TS game, each player can be a leader or a follower as the case may be. Therefore, the position of game players is dynamically transformable according to current conditions.

TS game model is a useful framework for designing decentralized mechanisms, such that users in SC systems can self-organize into the mutually satisfactory resource sharing process. This self-organizing feature can add autonomics into SC systems and help to ease the heavy burden of complex centralized control algorithms. Especially, the DSCM scheme pays serious attention to trust evaluation, repeated interactions and iterative self-learning techniques to effectively implement the resource sharing process. In the DSCM scheme, such techniques have been incorporated into the TS game model, and work together toward an effective system performance.

Transformable Stackelberg Game Model

Social cloud is a form of community cloud and is designed to enable access to elastic compute capabilities contributed by socially connected community (Caton, 2014). To avoid the social dilemma such as '*Tragedy of the Commons*', social incentives motivate users to participate in, and contribute to, SC systems in different ways. Motivation is generally categorized as either intrinsic or extrinsic. Extrinsic motivation represents that users are motivated by an external reward, e.g., virtual currency. Therefore, they will contribute to the SC while the expected benefits exceed the cost of contribution. Intrinsic motivation represents an internal satisfaction obtained from the task itself rather than the rewards or benefits. In realities, people incline to cooperate with others for reciprocation and altruism. These factors rationalize non-economic behaviors and motivates users to contribute to SC (Chard, 2012).

The DSCM scheme leverages social incentives to create ad hoc clouds without incurring the overhead of central complex processes. In the DSCM scheme, these techniques have been incorporated into the TS game model, which is developed to let distributed players learn the best strategy in the step-by-step interactive online manner. This approach can induce all SC users to share resources as much as possible and ensure a good tradeoff between the implementation complexity for real-world SC operations and an effective system performance. Therefore, the DSCM scheme can be used to overcome one of the major limitations of traditional SC monitoring methods.

In a realistic SC scenario, each user, i.e., network device, can be a resource supplier or demander. Suppliers make their decisions by considering the possible reactions of demanders. Demanders react dependently based on the decision of suppliers while attempting to maximize their satisfaction. Therefore, in the TS game model ($\mathbb{G}$), suppliers plays the role of leaders and demanders become followers. Based on these assumptions, $\mathbb{G}$ is defined as a tuple $\mathbb{G}$ =

$$\left(\mathbb{N},\left(V_i\right)_{i\in\mathbb{N}},\left(\boldsymbol{S}_i\right)_{i\in\mathbb{N}},\left(\Lambda_i\right)_{i\in\mathbb{N}},\left(U_i\right)_{i\in\mathbb{N}},T\right)$$

at each time period t of gameplay.

- $\mathbb{N}$ is the finite set of players, $\mathbb{N} = \{p_1, \ldots, p_n\}$ where $p_{i,1 \le i \le n}$ represents the ith user. A player can be a supplier or a demander at times. Therefore, the position of each player would be dynamically changeable as a leader or a follower.
- V_i is the amount of exchangeable resources of the player i. In this scheme, V is the computing capacity, e.g. CPU cycles.
- $\boldsymbol{S}_i$ is the set of strategies with the player i. If the player i is a supplier, $\boldsymbol{S}_i$ can be defined as the amount of sharing resource. If the player i is a demander, $\boldsymbol{S}_i$ is defined as the amount of requested resource.
- Λ_i is the contribution level of the player i in the SC community.
- The U_i is the payoff received by the player i. Traditionally, the payoff is determined as the obtained outcome minus the cost to obtain that outcome.
- The T is a time period. The $\mathbb{G}$ is repeated $t \in T < \infty$ time periods with competitive and cooperative manner.

In the SC system, each network device has its own computation resources and executes elastic applications. Applications can be divided into two parts: one part runs locally and the other part can be executed on the cloud side. Therefore, applications in each network device can be computed either locally or remotely via computation offloading. In general, the main challenges to design an offloading mechanism are to decide what, when and how to be offloaded. In the DSCM scheme, available resources in suppliers are matched to demanders based on the supplier-demander interactive relationship. According to the TS game model, network devices can self-organize into the mutually satisfactory computation offloading decisions.

Resource Sharing Process in Social Cloud Systems

Different users may pursue individually to maximize their profits. From the viewpoint of demanders, the payoff corresponds to the resource sharing benefit minus the incurred cost to share the remote resource. Therefore, the utility function of demander i (U_i^D) is defined as follows:

$$U_i^D\left(x_i, \Lambda_i\right) = \mathcal{B}_i\left(j, x_i\right) - \mathcal{C}_i(j, x_i), \text{ s.t., } j \text{ is a supplier} \in \mathbb{N} \text{ and } i \neq j \tag{35}$$

where x_i is the requested resource amount, and $\mathcal{B}_i(\cdot)$ and $\mathcal{C}_i(\cdot)$ are the benefit and cost functions for the demander i. Usually, elastic applications have concave benefit function, which provides monotone increasing values in proportion to the assigned resource amounts. According to the amount of assigned resource, $\mathcal{B}_i(\cdot)$ and $\mathcal{C}_i(\cdot)$ are given by

$$\mathcal{B}_i(j, x_i) = \sin\left(\frac{\pi}{2} \times \frac{b_j^i}{x_i}\right) \text{ and } \mathcal{C}_i(j, x_i) = \zeta \times \left(\varrho \times \frac{b_j^i}{\mathcal{MX}}\right)$$

s.t.,

$$\zeta = {b_j^i} \Big/ {\max\{\Lambda_i, b_j^i\}} \text{ and } \varrho = {\varepsilon(b_j^i)} \Big/ {\varepsilon(\mathcal{MX})} \tag{36}$$

where b_j^i is the assigned resource amount from the supplier j. ζ is a cost control parameter, and $\mathcal{MX}$ is the total resource amount to process the corresponding application. $\varepsilon(\mathcal{MX})$ and $\varepsilon(b_j^i)$ are the energy consumption to execute $\mathcal{MX}$ and b_j^i amount resources, respectively. In the DSCM scheme, $\varepsilon(\cdot)$ is a linear function. Λ_i is the accumulated contributiveness of the demander i. After the remote execution, Λ_i is decreased by b_j^i, i.e., $\Lambda_i = \Lambda_i - b_j^i$. Based on the expected payoff $U^D(\cdot)$, demanders can try to find the best actions, i.e., the decision of x_i amount.

From the viewpoint of suppliers, the payoff also corresponds to the received benefit minus the incurred cost to assign the sharing resource. However, in contrast to the demanders' interest, the sharing benefit is defined according to the reciprocal cooperation, more generally, the combination of evolution, altruism, and reciprocity. In the DSCM scheme, users can be altruistic toward others and react to other users' altruism. Therefore, the received benefit function is developed based on the simple reciprocal mechanism. By considering the service cost, the supplier j's utility function to the demander i ($U_j^S(\cdot)$) is defined as follows:

$$U_j^S(\mathcal{Z}_j, \Lambda_j, i) = \mathbb{B}_j(\mathcal{Z}_j, \Lambda_j, i) - \mathbb{C}_j(\mathcal{Z}_j) \tag{37}$$

where $\mathcal{Z}_j$ is the amount of sharing resource of the supplier j, and $\mathbb{B}_j(\mathcal{Z}_j, \Lambda_j, i)$ and $\mathbb{C}_j(\mathcal{Z}_j)$ are the benefit and cost functions for the supplier j, respectively. To get the optimal payoff, suppliers try to maximize their benefit function while minimizing their cost function. According to the $\mathcal{Z}_j$ and Λ values, $\mathbb{B}_j(\cdot)$ and $\mathbb{C}_j(\cdot)$ are given by.

$$\mathbb{B}_j(\mathcal{Z}_j, \Lambda_j, i) = \left[\left(\theta_j^i \times e^{\mathcal{Z}_j}\right) + \mathcal{F}_j(\Lambda_j)\right] \text{and } \mathbb{C}_j(\mathcal{Z}_j) = \lambda \times \left(\varepsilon(\mathcal{Z}_j) / \varepsilon(\mathfrak{T})\right)$$

s.t.,

$$\theta_j^i = \frac{\varphi_j + \left(\varphi_j \times \left(\Lambda_i / (\Lambda_i + \Lambda_j)\right)\right)}{1 - \left(\Lambda_i / (\Lambda_i + \Lambda_j)\right)} \text{ and } \mathcal{F}_j(\Lambda_j) = \left(\mathcal{Z}_j / \max\{\Lambda_j, \mathcal{Z}_j\}\right) \quad (38)$$

θ_j^i is the supplier j's altruistic parameter to the demander i, and φ_j is the supplier j's general altruistic propensity. $\varepsilon(\mathfrak{T})$ and $\varepsilon(\mathcal{Z}_j)$ are the energy consumption to execute the supplier j's total resource ($\mathfrak{T}$) and $\mathcal{Z}_j$, respectively. λ is the cost control parameter. After the resource sharing process, Λ_j is increased by $\mathcal{Z}_j$, i.e., $\Lambda_j = \Lambda_j + \mathcal{Z}_j$.

Under dynamically changing SC environments, a fixed altruistic propensity cannot effectively adapt to the current SC condition. Therefore, the φ value should be dynamically adjustable. In order to implement the φ value adjustment process, suppliers should learn how to perform well by interacting with demanders and dynamically adjust their φ levels. Based on the exponential weight learning algorithm (Gajane, 2015), suppliers in the TS game model can constantly adapt each φ level to get an appropriate attitude to their corresponding SC environments. Let $\mathbb{K}$ be the set of all possible altruistic propensity levels, i.e., $\varphi \in \mathbb{K}$. In the DSCM scheme, the probability of choosing the k's propensity level in $\mathbb{K}$ at time t ($P_k^{\varphi}(t)$) is defined by.

$$P_k^{\varphi}(t) = (1-\gamma) \times \left(\omega_k(t) \Big/ \sum_{j=1}^{K} \omega_j(t) \right) + \frac{\gamma}{\|\mathbb{K}\|}$$

s.t.,

$$\omega_j(t) = \omega_j(t-1) \times \exp\left(\gamma \times \left[\mathcal{U}_j(t-1) \Big/ \left(P_j^{\varphi}(t-1) \times \|\mathbb{K}\| \right) \right] \right) \quad (39)$$

where $\gamma \in [0,1]$ is an egalitarianism factor, which tunes the desire to pick an action uniformly at random. That is, if $\gamma = 1$, the weights have no effect on the choices at any step. $\|\mathbb{K}\|$ is the total number of propensity levels, and $\mathcal{U}_j(t-1)$ is the obtained payoff ($U_j^S(\cdot)$) at time $t-1$. According to the distribution of $P(t)$, suppliers can modify their φ levels without any impractical rationality assumptions. During the step-by-step iteration, suppliers individually adjust the φ value by using the dynamics of feedback-based repeated process. Therefore, under dynamic SC situations, the main advantage of DSCM scheme is a real-world practicality.

During real-world SC operations, multiple demanders can request the resource sharing from the same supplier. In this case, the role of supplier is to distribute dynamically the limited resource for each demander. To get a fair-efficient resource allocation, the DSCM scheme develops a new resource distribution algorithm based on the relative utilitarian bargaining model (Kim, 2014); it can be applicable and useful in a SC system with a frequently changing situation. In the DSCM scheme, demanders' Λ values are considered as asymmetric bargaining powers. Therefore, the bargaining solution ($\mathcal{R}\_\mathcal{B}$) for resource distribution is given by.

$$\mathcal{R}\_\mathcal{B} = \max_{b_j^i,\ i \in \mathcal{N}_j} \left(\sum_{i \in \mathcal{N}_j} \mathfrak{U}_i \left(b_j^i, \Lambda_i \right) \right), s.t., \mathfrak{U}_i \left(b_j^i, \Lambda_i \right) = \left(\frac{b_j^i}{x_i} \right)^{\eta_i} \text{ and } \eta_i = \Lambda_i \Big/ \sum_{j=1}^{K} \Lambda_j \quad (40)$$

where $\mathcal{N}_j$ is the set of all resource requesting demanders to the supplier j. $\mathcal{R}\_\mathcal{B}$ is a vector, which corresponds to the resource distribution amounts to each demander.

In general, traditional game models have focused on investigating *which* decisions are made or *what* decisions should be made. Therefore, an equilibrium point is a well-known solution concept in classical game models. The strategy in equilibrium is the best response to the strategies of the other users. In the TS game model, an equilibrium point of suppliers and demanders are can be defined as follows:

$$U^*\left(U^{D*},U^{S*}\right)=$$
$$\begin{cases}\begin{cases}U^{S*}=\arg\max\limits_{\mathcal{Z}_j\in S_j}\left\{U_j^S\left(\mathcal{Z}_j,\Lambda_j,i\right)\right\}, & \textit{if } j \textit{ is a supplier with single demander } i\\ \arg\max\limits_{i\in\mathcal{N},\mathcal{Z}_j^i\in S_j}\left\{U_j^S\left(\mathcal{Z}_j,\Lambda_j,\mathcal{N}\right)\right\}, & \textit{if } j \textit{ is a supplier with multiple demanders } N\end{cases}\\ U^{D*}=\arg\max\limits_{x_i\in S_i}\left\{U_i^D\left(x_i,\Lambda_i\right)\right\}, \quad \textit{if } i \textit{ is a demander}\end{cases} \tag{41}$$

In recent decades, there had been many conceptual and empirical critiques toward the equilibrium concept. First, in the scenario of equilibrium, the players are assumed to be fully rational. This perfect rational assumption requires complete information; all factors of the game should be common knowledge. However, in reality, this assumption is actually disputable, and rarely holds. In particular, the hypothesis of exact rationality does not apply to many interactive situations. Second, the idea of equilibrium has mostly been developed in a static setting. Under the dynamic changing SC environments, it cannot capture the adaptation of players to change their strategies and reach equilibrium over time.

The DSCM scheme introduces a new solution concept for the TS game model; it is the obtained consensus with reciprocal advantage. Such a consensus in multi-player decision making process is defined as *Cooperative Consensus Equilibrium* (*CCE*). During TS game operations, game players may adjust their altruistic propensities when outcomes contradicts their beliefs, and adaptively modify their altruistic propensities in an attempt to reach a mutually acceptable decision vector. Therefore, the solution concept of *CCE* presents a dynamic learning interpretation to adapt the current SC situations.

Definition: *CCE* is a system status that can be obtained through repeating the TS game with receiving feedbacks. When all the accumulated contributiveness (Λ) of users are balanced, i.e., the relative contribution differences of users are less than a pre-defined maximum bound (Γ_Λ), this state is defined as the *CCE*. That is formally formulated as

$$\max_{i}\left\{i \in \mathbb{N} \,\middle|\, \left(\left(\Lambda_i / T_i^M\right) \Big/ \left(\sum_{k \in \mathbb{N}} \Lambda_k / T_k^M\right)\right)\right\} < \Gamma_\Lambda \tag{42}$$

where T_k^M is the maximum resource capacity of user k's device. Therefore, the main idea of *CCE* is to minimize the maximum unbalanced behavior degree of users.

The Main Steps of the DSCM Scheme

The DSCM scheme present a new TS game model for the interaction of multiple users with elastic applications. In the TS game, a sophisticated combination of the reciprocal relationship and incentive mechanism can provide much more suitable resource sharing algorithm. Based on the real-time interactive feedback process, each user can adapt its behavior and act strategically to achieve a better profit. The DSCM scheme is described by the following major steps.

Step 1: At the start, all Λ values are set to the relatively same initial values, e.g., zero, and each altruistic propensity φ is randomly chosen from $\mathbb{K}$. When reciprocal interaction history is unavailable, it is a proper initialization. Control parameters, i.e., γ, λ and Γ_Λ, are listed in Table1.

Step 2: When an individual device needs an additional resource, it becomes a demander, and asks the x amount resource to maximize the expected payoff $U^D(\cdot)$ according to equation (35).

Step 3: If the neighboring nodes of a demander have enough available resources, they can be suppliers. Suppliers provide the $\mathcal{Z}$ amount resource to maximize the expected payoff $U^S(\cdot)$ according to equation (37). When multiple demanders request the resource simultaneously, a supplier distributes the available resource using the equation (40).

Step 4: Using the simple two-sided matching algorithm, a demander selects the most adaptable supplier, and the resource is effectively shared. After the resource sharing process, Λ is adjusted dynamically.

Step 5: In each game stage game, φ of each mobile device is periodically modified according to the exponential weight learning algorithm. Based on the adjusted $P^{\varphi}_{k,k\in\mathbb{K}}$ in equation (39), an actual φ value of each device is selected stochastically.

Step 6: As game players, mobile devices are interrelated with elastic applications, and interact with each other in the TS game. Under widely diverse SC environments, this iterative feedback procedure continues to reach the *CCE* status.

Step 7: Mobile devices self-monitors the current SC situation in a distributed online manner; the next iteration resumes at Step 2.

Summary

The ever increasing use of social networks and arrival of new computing paradigms like cloud computing has urged the need to integrate these platforms for the better and inexpensive usage of resources. Sharing cloud resources in such environments would be very helpful. The DSCM scheme addresses a new resource control algorithm for SC systems. Using the TS game model, users iteratively observed the received payoffs and repeatedly modified their altruistic propensities to effectively manage SC resources. The DSCM scheme enables the sharing of SC resources between users via reciprocal cooperative relationships, and can effectively approach the *CCE* status using a step-by-step feedback process.

EMBEDDED GAME BASED FOG COMPUTING (EGFC) SCHEME

The *Embedded Game based Fog Computing* (EGFC) scheme is developed as a novel Fog Radio Access Networks (F-RAN) system control scheme based on the embedded game model. In the EGFC scheme, spectrum allocation, cache placement and service admission algorithms are jointly designed to maximize system efficiency. By developing a new embedded game methodology, the EGFC scheme can capture the dynamics of F-RAN system and effectively compromises the centralized optimality with decentralized distribution intelligence for the faster and less complex decision making process.

Development Motivation

In the past decade, the evolution toward 5G is featured by the explosive growth of traffic in the wireless network, due to the exponentially increased number of user devices. Compared to the 4G communication system, the 5G system should bring billions of user devices into wireless networks to demand high bandwidth connections. Therefore, system capacity and energy efficiency should be improved to get the great success of 5G communications. Cloud Radio Access Network (C-RAN) is an emerging architecture for the 5G wireless system. A key advantage of C-RAN is the possibility to perform cooperative transmissions across multiple edge nodes for the centralized cloud processing. However, the cloud processing comes at the cost of the potentially large delay entailed by fronthaul transmissions. It may become a major performance bottleneck of a C-RAN system per critical indicators such as spectral efficiency and latency (Hung, 2015; Park, 2016; Tandon, 2016).

As an extension of C-RAN paradigm, fog computing is a promising solution to the mission critical tasks involving quick decision making and fast response. It is a distributed paradigm that provides cloud-like services to the network edge nodes. Instead of using the remoted cloud center, the fog computing technique leverages computing resources at the edge of networks based on the decentralized transmission strategies. Therefore, it can help overcome the resource contention and increasing latency. Due to the effective coordination of geographically distributed edge nodes, the fog computing approach can meet the 5G application constraints, i.e., location awareness, low latency, and supports for mobility or geographical distribution of services. The most frequently referred use cases for the fog computing concept are related to the IoT (Borylo, 2016; Dastjerdi, 2016).

Taking full advantage of fog computing and C-RANs, Fog Radio Access Networks (F-RAN) has been proposed as an advanced socially-aware mobile networking architecture in 5G systems. F-RANs harness the benefits of, and the synergies between, fog computing and C-RAN in order to accommodate the broad range of Quality of Service (QoS) requirements of 5G mobile broadband communication (Tandon, 2016a). In the F-RAN architecture, edge nodes may be endowed with caching capabilities to serve the local data requests of popular content with low latency. At the same time, a central cloud processor allocates radio and computational resources to each individual edge nodes while ensuring as much as various applications (Tandon, 2016). To maximize the F-RAN system performance, application request scheduling, cache placement, and communication resource allocation should be jointly designed. However, it is an extremely challenging issue.

In the architecture of F-RANs, multiple interest relevant system agents exist; they are the central Cloud Server (CS), Edge Nodes (ENs) and Mobile Users (MUs). The CS provides contents to download, allocates radio and communication resources to ENs. ENs, known as Fog-computing based Access Points (F-APs), manage the allocated radio resource, and admit MUs to provide application services. MUs wish to enjoy different QoS services from the F-RAN system. Different system agents have their own benefits, but their benefits could conflict with each other, and each agent only cares about its own profit. Therefore, it is necessary to analyze the interactions among these conflicting system agents and design proper solutions. Although dozens of techniques have been proposed, a systematic study on the interactions among CS, F-APs and MUs is still lacking (Hu, 2016).

The traditional game theoretic analysis should rely on the perfect information and idealistic behavior assumptions. Therefore, there is a quite general consensus to say that the predicted game solutions are useful but would be rarely observed in real world situations. Recently, specialized sub-branches of game theory have been developed to encounter this problem. To design a practical game model for the F-RAN system management, the EGFC scheme adopts an online dynamic approach based on the interactive relationship among system agents. This approach exploits a partial information on the game, and obtains an effective solution under mild and practical assumptions. From the standpoint of algorithm designers, this approach can be dynamically implemented in the real-world F-RAN environments.

Embedded Game Model for F-RAN Systems

In the C-RAN architecture, all control functions and application storage are centralized at the CS, which requires a lot of MUs to transmit and exchange their data fast enough through the fronthaul link. To overcome this C-RAN's disadvantage with the fronthaul constraints, much attention has been paid to mobile fog computing and the edge cloud. The design of fog computing platform has been introduced to deliver large-scale latency-sensitive applications. To implement the fog computing architecture, traditional edge nodes are evolved to the Fog-computing based Access Point (F-AP) by being equipped with a certain caching, cooperative radio resource and computation power capability (Peng, 2016; Tandon, 2016a).

The main difference between the C-RAN and the F-CRAN is that centralized storage cloud and control cloud functions are distributed to individual F-APs. Usually, F-APs are used to forward and process the received data, and interface to the CS through the fronthaul links. To avoid all traffic being loaded directly to the centralized CS, some local traffic should be delivered from the caching located in F-APs. Therefore, each F-AP integrates not only the front radio spectrum, but also the locally distributed cached contents and computation capacity. This approach can save the spectral usage of constrained fronthaul while decreasing the transmission delay. In conclusion, the main characteristics of F-RAN include ubiquity, decentralized management and cooperation (Peng, 2016; Tandon, 2016a).

During the F-RAN system operations, system agents, i.e., CS, F-APs, MUs - should make decisions individually by considering the mutual-interaction relationship. Under the dynamic F-RAN environments, system agents try to maximize their own profits in a competitive or cooperative manner. The EGFC scheme develops a new game model, called embedded game, for the F-RAN system. According to the decision making method, the embedded game procedure can be divided two phases. At the first phase, the CS and F-APs play a superordinated game; the CS distribute the available spectrum resource to each F-APs by using a cooperative manner. At the second phase, F-APs and MUs play subordinated games. By employing a non-cooperative manner, an individual F-AP selectively admits its corresponding MUs to provide different application services. Taken as a whole, multiple subordinated games are nested in the superordinated game.

Formally, the EGFC scheme defines the embedded game model $\mathbb{G}=\{\mathbb{G}^{super}, \mathbb{G}^{sub}_{i,1\leq i\leq n}\}$ where $\mathbb{G}^{super}$ is a superordinated game to formulate interactions between CS and F-APs, and $\mathbb{G}^{sub}_{i}$ is a subordinated game to formulate interactions between the ith F-AP and its corresponding MUs. Firstly, the $\mathbb{G}^{super}$ can be defined as

$$\mathbb{G}^{super} = \left\{\mathbb{N}, \mathcal{R}_{CS}, \boldsymbol{S}^{\mathcal{R}}_{CS}, U_{i,1\leq i\leq n}, T\right\}$$

at each time period t of gameplay.

- $\mathbb{N}$ is the finite set of $\mathbb{G}^{super}$ game players

$$\mathbb{N}=\{\text{CS}, F\text{-}AP_1, F\text{-}AP_2 \ldots F\text{-}AP_n\}$$

where the total $n+1$ number of $\mathbb{G}^{super}$ players; one CS and n F-APs.

- The total spectrum resources of CS is $\mathcal{R}_{CS}$, which would be distributed to n F-APs.
- $\boldsymbol{S}^{\mathcal{R}}_{cs} = \{\delta_1, \delta_2, \ldots \delta_n\}$ is the sets of CS's strategies for the spectrum resource allocation. δ_i in $\boldsymbol{S}^{\mathcal{R}}_{cs}$ is the allocated spectrum amount for the $F\text{-}AP_{i,1\leq i\leq n}$.

- The $U_{i,1\le i\le n}$ is the payoff received by the $F\text{-}AP_i$. It is estimated as the obtained outcome minus the cost from the spectrum resource allocation.
- The T is a time period. The $\mathbb{G}^{super}$ is repeated $t \in T < \infty$ time periods with imperfect information.

Secondly, the $\mathbb{G}_i^{sub}$ is the ith subordinated game, and it can be defined as

$$\mathbb{G}_i^{sub} = \left\{\mathbb{M}_i, \Re_i, \boldsymbol{S}_{F-AP_i}^{\delta_i}, \boldsymbol{S}_{F-AP_i}^{\mathcal{C}^i}, \boldsymbol{S}_{F-AP_i}^{\sigma^i}, \mathcal{U}_{j,1\le j\le m}^i, T\right\}$$

at each time period t of gameplay.

- $\mathbb{M}_i$ is the finite set of $\mathbb{G}_i^{sub}$ game players

$$\mathbb{M}_i = \{F\text{-}AP_i, MU_1^i, \ldots, MU_m^i\}$$

where $MU_{j,1\le j\le m}^i$ is the jth MU in the area covered by the $F\text{-}AP_i$.

- The set of $F\text{-}AP_i$'s resources is $\Re_i = \{\delta_i, \mathcal{C}_i, \sigma_i\}$ where δ_i, $\mathcal{C}_i$, σ_i are the allocated spectrum resource, the computation capacity, and the placed cache files in the $F\text{-}AP_i$, respectively.
- $\boldsymbol{S}_{F-AP_i}^{\delta_i}$, $\boldsymbol{S}_{F-AP_i}^{\mathcal{C}_i}$ and $\boldsymbol{S}_{F-AP_i}^{\sigma_i}$ are the sets of $F\text{-}AP_i$'s strategies for the spectrum allocation for MUs, the computation capacity assignment for MUs, and cache placement in the $F\text{-}AP_i$, respectively.
- The $\mathcal{U}_{j,1\le j\le m}^i$ is the payoff received by the $F\text{-}AP_i$.
- The T is a time period. The $\mathbb{G}_i^{sub}$ is repeated $t \in T < \infty$ time periods with imperfect information.

Solution Concept for the Superordinated Game

In the superordinated game, game players are CS and F-APs, and they are rational to reach a win-win situation. In many situations, each rational agent is able to improve his objectives without preventing others from improving their objectives. Therefore, they are more prone to coordinate and willing to play cooperative games (Qiao, 2006). Usually, solution concepts are different in different games. For the CS and F-APs interactions, the *Kalai and Smoro-*

dinsky Bargaining Solution (KSBS) is an interesting solution concept. Like as the well-known *Nash Bargaining Solution* (NBS), the KSBS also provides a fair and optimal solution in a cooperative manner. In addition, the KSBS can be used when the feasible payoff set is not convex. It is the main advantage of KSBS over the NBS. Due to this appealing property, the KSBS approach has been practically implemented to solve real-world problems (Kim, 2014).

In order to show the effectiveness of the KSBS, it is necessary to evaluate each player's credibility. The EGFC scheme obtains the KSBS based on the F-APs' trustworthiness. This information can be inferred implicitly from the F-APs' outcome records. Therefore, the EGFC scheme can enhance the effectiveness of KSBS while restricting the socially uncooperative F-APs. At time t, the $F\text{-}AP_i$'s trust assessment $\left(\mathcal{T}_i(t)\right)$ for the spectrum allocation process is denoted by

$$\mathcal{T}_i(t) = \left\{(1-\beta)\times\mathcal{T}_i(t-\Delta t)\right\} + \left\{\beta \times \left(\left(\frac{U_i(\Delta t)}{\sum_{j=1}^{n} U_j(\Delta t)}\right) \Big/ \left(\frac{\delta_i(t-\Delta t)}{\mathcal{R}_{CS}}\right)\right)\right\}$$

s.t.,

$$\beta = \left(\phi\times\mathcal{T}_i(t)\right) \Big/ \left(1+\left\{\phi\times\mathcal{T}_i(t)\right\}\right) \text{ and } \phi \geq 0 \tag{43}$$

where $U_i(\Delta t)$ is the throughput of the $F\text{-}AP_i$ during the recent Δt time period, and $\delta_i(t-\Delta t)$ is the δ_i value at the time period $[t-\Delta t]$. The parameter β is used to weigh the past experience by considering a trust decay over time. In addition, the EGFC scheme introduces another parameter ϕ to specify the impact of past experience on $\mathcal{T}_i(t-\Delta t)$. Essentially, the contribution of current information increases proportionally as ϕ increases. In this case, the EGFC scheme can effectively adapt to the currently changing conditions while improving resiliency against credibility fluctuations (Bao, 2012).

Under the dynamic F-RAN environment, F-APs request individually their spectrum resource to the CS at each time period. To adaptively respond the current F-RAN system conditions, the sequential KSBS bargaining approach gets the different KSBS at each time period. It can adapt the timely dynamic F-RAN situations. At time t, the timed KSBS ($\mathfrak{F}_{KSBS}^{t}$) for the spectrum resource problem is mathematically defined as;

$$\mathfrak{F}_{KSBS}^{t}\left(\boldsymbol{S}_{cs}^{\mathcal{R}}\right)=\left\{\delta_1(t),\delta_2(t),\dots\delta_n(t)\right\}=$$

$$\left(\frac{\sup\left\{U_1^t\left(\delta_1(t)\right)\right\}-d_1}{\omega_1^t\times\left(\mathbb{O}_1^t-d_1\right)}=\dots=\frac{\sup\left\{U_i^t\left(\delta_i(t)\right)\right\}-d_i}{\omega_i^t\times\left(\mathbb{O}_i^t-d_i\right)}=\dots=\frac{\sup\left\{U_n^t\left(\delta_n(t)\right)\right\}-d_n}{\omega_n^t\times\left(\mathbb{O}_n^t-d_n\right)}\right)$$

s.t.,

$$\mathbb{O}_i^t=\max\left\{U_i^t\left(\delta_i(t)\right)\mid U_i^t\left(\delta_i(t)\right)\in\mathbb{R}^n\right\},\ \omega_i^t=T_i(t)\Big/\sum_{j=1}^{n}T_j(t)$$

and

$$\sup\left\{U_i^t\left(\delta_i(t)\right)\right\}=\sup\left\{U_i^t\left(\delta_i(t)\right):\left\{\left(U_i^t\left(\delta_1(t)\right),\dots,U_n^t\left(\delta_n(t)\right)\right)\right\}\subset\mathbb{R}^n\right\}\qquad(44)$$

where $U_i^t\left(\delta_i(t)\right)$ is the $F\text{-}AP_i$'s payoff with the strategy δ_i during the recent time period (Δt). $\mathbb{R}^n$ is a jointly feasible utility solution set, and a disagreement point (***d***) is an action vector $\boldsymbol{d}=(d_1,..\ d_n)\in\mathbb{R}^n$ that is expected to be the result if players, i.e., F-APs, cannot reach an agreement (i.e., zero in the system). ω_i^t ($0<\omega_i^t<1$) is the player $F\text{-}AP_i$'s bargaining power at time t, which is the relative ability to exert influence over other players. $\mathbb{O}_i^t$ is the ideal point of player $F\text{-}AP_i$ at time t. Therefore, players choose the best outcome subject to the condition that their proportional part of the excess over the disagreement is relative to the proportion of the excess of their ideal gains. Simply, the EGFC scheme can think that the KSBS is the intersection point between the Pareto boundary and the line connecting the disagreement to the ideal gains (Kim 2016). Therefore, in the EGFC scheme,

$$\boldsymbol{S}_{cs}^{\mathcal{R}} = \mathfrak{F}_{KSBS}^{t}\left(\boldsymbol{S}_{cs}^{\mathcal{R}}\right) = \left\{\delta_1\left(t\right), \delta_2\left(t\right), \ldots \delta_n\left(t\right)\right\}$$

is a joint strategy, which is taken by the CS at time t.

In non-deterministic settings, $\mathfrak{F}_{KSBS}^{t}\left(\boldsymbol{S}_{cs}^{\mathcal{R}}\right)$ is a selection function to define a specific spectrum allocation strategy for every F-APs. Due to the main feature of KSBS, the increasing of bargaining set size in a direction favorable to a specific F-AP always benefits that F-AP. Therefore, in the superordinated game, self-interested F-AP can be satisfied during the F-RAN system operations. To practically obtain the $\mathfrak{F}_{KSBS}^{t}\left(\boldsymbol{S}_{cs}^{\mathcal{R}}\right)$ in the equation (44), the EGFC scheme can re-think the KSBS as a weighted max-min solution like as;

$$\mathfrak{F}_{KSBS}^{t}\left(\boldsymbol{S}_{cs}^{\mathcal{R}}\right) = \left\{\delta_1\left(t\right), \delta_2\left(t\right), \ldots \delta_n\left(t\right)\right\} = \arg \max_{\left\{\delta_1(t), \delta_2(t), \ldots \delta_n(t)\right\}} \left\{ \min_{\delta_{i,1\leq i\leq n}(t)} \left(\frac{\sup\left\{U_i^t\left(\delta_i\left(t\right)\right)\right\} - d_i}{\omega_i^t \times \left(\mathbb{O}_i^t - d_i\right)} \right) \right\} \quad (45)$$

Solution Concept for the Subordinated Games

Edge processing is the key emerging trends in the F-RAN system. It refers to the localization of computing, communication, and storage resources at the F-APs. In the F-RAN architecture, F-APs are connected to the CS through fronthaul links. Under this centralized structure, the performance of F-RANs is clearly constrained by the fronthaul link capacity; it incurs a high burden on fronthaul links. Therefore, a prerequisite requirement for the centralized CS processing is the high bandwidth and low latency fronthaul interconnections. However, during the operation of F-RAN system, unexpected growth of service requests may create a traffic congestion. It has a significant impact on the F-RAN performance. To overcome the disadvantages of F-RAN architecture imposed by the fronthaul constraints, new techniques have been introduced with the aim of reducing the delivery latency by limiting the need to communicate between the CS and MUs (Tandon, 2016b).

Currently, there are evidences that MUs' downloading of on-demand multimedia data is the major reason for the data avalanche over F-RAN; numerous repetitive requests on the same data lead to redundant transmissions. Usually, multimedia data are located in the CS and far away from MUs. To ensure an excellent QoS provisioning, an efficient solution is to locally store these frequently-access data into the cache memory of F-APs while reducing the transmission latency; it is known as caching. This approach can effectively mitigate the unnecessary fronthaul overhead caused by MUs' repetitive service requests. Therefore, CS, F-APs and MUs are all the beneficiaries from the local caching mechanism (Li, 2016).

In the subordinated game, an efficient caching mechanism is designed by carefully considering the relations and interactions among CS, F-APs and MUs. Therefore, it not only can the heavy traffic load be relieved at fronthaul links, but also the request latency can be decreased, which results in better QoS (Tandon, 2016b). A practical caching mechanism is coupled with the data placement. In the F-RAN architecture, the EGFC scheme assumes that a multimedia file set $\mathbb{M} = \{\mathcal{M}_1, \ldots, \mathcal{M}_L\}$ consists of L popular multimedia files in the CS, and files in $\mathbb{M}$ can be possibly cached in each F-AP. The popularity distribution among $\mathbb{M}$ is represented by a vector $\mathcal{Q} = [g_1, \ldots, g_L]$. Generally, the vector $\mathcal{Q}$ can be modeled by a Zipf distribution (Li, 2016);

$$g_l = \left(\frac{1}{l^{\tau}}\right) \Big/ \left(\sum_{f=1}^{L} \frac{1}{f^{\tau}}\right), \text{ s.t., } 1 \le l \le L \text{ and } \tau > 0 \tag{46}$$

where τ factor characterizes the file popularity. In the EGFC scheme, MUs in each F-AP area are assumed to request independently the lth file $\mathcal{M}_{l,1\le l\le L}$. Therefore, the τ value is different for each F-AP. According to (46), $\mathcal{M}_1$ (*or* $\mathcal{M}_L$) has the highest (*or* lowest) popularity. The CS intends to rent a frequency-accessing fraction of $\mathbb{M}$ for caching to maximize the F-RAN system performance. The EGFC scheme denotes the caching placement strategy as a two-dimensional matrix $\mathbb{I} = [0,1]^{n\times L}$ consisting of binary entries where 1 is indicating the caching placement in a F-AP, and 0 is not. $\mathbb{I}$ is defined as

$$\mathbb{I} \triangleq \begin{bmatrix} I_1^1 & \cdots & I_1^L \\ \vdots & \ddots & \vdots \\ I_n^1 & \cdots & I_n^L \end{bmatrix} \in [0,1]^{n\times L} \tag{47}$$

where $I_i^l = 1$ means that the file $\mathcal{M}_l$ is cached at the $F\text{-}AP_i$ and $I_i^l = 0$ means the opposite. For the $F\text{-}AP_i$, the profit ($\mathfrak{R}_i^c$) gained from the local caching mechanism can be defined as follows;

$$\mathfrak{R}_i^c = \sum_{l=1}^{L}\left(g_l^i \times \mathcal{L}^i \times \mathcal{Z}_l^i \times I_i^l\right) - \sum_{l=1}^{L}\left(\mathfrak{C}_l^i \times I_i^l\right), s.t., g_l^i \in \mathcal{Q}^i \tag{48}$$

where $\mathcal{Q}^i$ is the vector $\mathcal{Q}$ of $F\text{-}AP_i$ and $\mathcal{L}^i$ is the total number of service requests on average. $\mathcal{Z}_i^l$ and $\mathfrak{C}_i^l$ is the revenue and cost from the caching in the $F\text{-}AP_i$, respectively. From the viewpoint of $F\text{-}AP_i$, the fraction $[\,I_i^1 \dots I_i^L\,]$ of $\mathbb{I}$ ($\mathcal{Q}^i$) needs to be optimized for maximizing the $\mathfrak{R}_i^c$.

Based on the current caching placement, Service Admission Control (SAC) algorithm should be developed to make admission decisions to maximize their spectrum efficiency while maintaining a desirable overhead level. Especially, when the requested services are heavy, that is, the sum of the requested resource amount exceeds the currently available system capacity, the SAC comes into act whether to accept a new service request or not. Based on the acceptance condition, such as the current caching status and resource capacity, the SAC problem can be formulated as a joint optimization problem. In this problem, the EGFC scheme takes into account the maximization of spectrum efficiency while minimizing the fronthaul overhead.

The EGFC scheme set out to obtain fundamental insights into the SAC problem by means of a game theoretic approach. Therefore, the subordinated game is designed to formulate the interactions of the F-AP and MUs while investigating the system dynamics with imperfect information. To implement the subordinated game, the EGFC scheme adopts the concept of dictator game, which is a game in experimental economics, similar to the ultimatum game, first developed by D. Kahneman et al (Daniel, 1986). In the dictator game, one player, called *the proposer*, distributes his resource, and the other players, called *the responders*, simply accept the decision, which is made by *the proposer*. As one of decision theory, the dictator game is treated as an exceptional non-cooperative game or multi-agent system game that has a partner-feature and involves a trade-off between self- and other-utility. Based on its simplicity, the dictator game can capture an essential characteristic of the repeated interaction situation (Kim, 2014).

In the subordinated game model, each F-AP is *the proposer* and MUs are *the responders*. They interact with each other and repeatedly work together toward an appropriate F-RAN performance. To effectively make SAC decisions, *the proposer* considers the current system conditions such as the available spectrum amount, the current caching placement and fronthaul overhead status. By a sophisticated combination of these conflicting condition factors, *the proposer* attempts to approximate a temporary optimal SAC decision. The SAC decision procedure is shown in Figure 2.

According to the SAC procedure, each $F\text{-}AP_i$ can maintain the finest SAC solution while avoiding the heavy computational complexity or overheads. For the subordinated game, the EGFC scheme proposes a new solution con-

Figure 2. Service admission control procedure

Define:

Θ_j^i : a new service request of MU_j^i,

$\boldsymbol{Min\_S}(\Theta_j^i)$, $\boldsymbol{Min\_C}(\Theta_j^i)$: the minimum spectrum and computation requirement of Θ_j^i

χ^i, y^i : the currently using spectrum and computation amount in the $F\text{-}AP_i$

X^i, M^i : the current and maximum fronthaul transmission rate

When the Θ_j^i is requested in the $F\text{-}AP_i$,

If Θ_j^i request is cached in the $F\text{-}AP_i$, // *no computation offloading task*

{ $\boldsymbol{If}\left(\chi^i + \boldsymbol{Min\_S}\left(\Theta_j^i\right) \le \boldsymbol{S}_{F-AP_i}^{P^i}\right)$, it is accepted

Otherwise, it is rejected }

Else {

If Θ_j^i is computation offloading task,

{ $\boldsymbol{If}\left(\left(\chi^i + \boldsymbol{Min\_S}\left(\Theta_j^i\right) \le \boldsymbol{S}_{F-AP_i}^{P^i}\right) \text{and} \left(y^i + \boldsymbol{Min\_C}\left(\Theta_j^i\right) \le \boldsymbol{S}_{F-AP_i}^{X^i}\right)\right)$,

it is accepted

Else if $\left(\left(\chi_j^i + \boldsymbol{Min\_S}\left(\Theta_j^i\right) \le \boldsymbol{S}_{F-AP_i}^{P^i}\right) \text{ and } \left(X^i + \boldsymbol{Min\_S}\left(\Theta_j^i\right) \le \eth \times M^i\right)\right)$

it is accepted,

Otherwise, it is rejected }

Else if $\left(\left(\chi_j^i + \boldsymbol{Min\_S}\left(\Theta_j^i\right) \le \boldsymbol{S}_{F-AP_i}^{P^i}\right) \text{ and } \left(X^i + \boldsymbol{Min\_S}\left(\Theta_j^i\right) \le \eth \times M^i\right)\right)$

it is accepted,

Otherwise, it is rejected

}

cept, *Temporal Equilibrium* (*TE*). In the EGFC scheme, all MUs follow compulsorily the decision of *F-APs*, and the outcome profile of SAC process constitutes the *TE*, which is the current service status.

$$TE = \overrightarrow{\mathcal{TE}}_i \mid \left(\mu_i \cup \psi_i\right) \to \left(\Theta^i_{j,1\le j\le m} \in \left(\mu_i \cup \psi_i\right)\right) \underline{\underline{def}}\ \overrightarrow{\mathcal{TE}}_i = \begin{cases} \Theta^i_j \in \mu_i, & if\ \Theta^i_j\ is\ accepted \\ \Theta^i_j \in \psi_i, & otherwise \end{cases} \tag{49}$$

The Main Steps of the EGFC Scheme

In the EGFC scheme, the superordinated game for spectrum allocation and the subordinated game for SAC decisions are interlocked and serially correlated. The subordinated game depends on the outcome of superordinated game, and the result of subordinated games is the input back to the superordinated game process. Structurally, the multiple subordinated games are nested in the superordinated game, and they are linked based on the step-by-step interactive feedback process. The main steps of the EGFC scheme are given next.

Step 1: At the initial time, the spectrum resource allocation $\boldsymbol{S}^{\mathcal{R}}_{cs} = \{\delta_1, \delta_2, \ldots. \delta_n\}$ and trustworthiness ($\mathcal{T}$) for F-APs are equally distributed. This starting guess guarantees that each F-AP enjoys the same benefit at the beginning of the game.

Step 2: Control parameters $\mathcal{C}$, n, m, σ, β, $\mathcal{R}_{CS}$, Δt, ϕ, $\mathcal{Z}$, $\mathfrak{C}$, τ, L, $\mathfrak{M}$ and ϵ are given from the simulation scenario (refer to the Table I).

Step 3: At each superordinated game period, $\boldsymbol{S}^{\mathcal{R}}_{cs} = \{\delta_1, \delta_2, \ldots. \delta_n\}$ is dynamically adjusted according to (43)-(45); it is the timed KSBS while taking into account the current F-RAN situations.

Step 4: The trustworthiness ($\mathcal{T}$) for each F-AP is modified periodically by using (43).

Step 5: In a distributed manner, the caching placement in each F-AP occurs while maximizing the $\mathfrak{R}^c$ according to the equation (48).

Step 6: Based on the assigned δ value, each F-AP performs a subordinated game. By considering the current system conditions such as χ, y, $\mathfrak{R}^c$ and $\mathfrak{X}$, the SAC procedure is executed in a real-time online manner.

Step 7: The superordinated and subordinated games are interlocked and serially correlated. Based on the interactive feedback mechanism, the dynamics of embedded game can cause cascade interactions of game players and players can make their decisions to quickly find the most profitable solution.

Step 8: Under widely diverse F-RAN environments, the CS and F-APs are self-monitoring constantly for the next embedded game process; proceed to Step 3.

Summary

As a promising paradigm for the 5G communication system, the F-RAN has been proposed as an advanced socially-aware wireless networking architecture to provide the higher spectral efficiency while maximizing the system performance. In the EGFC scheme, the SAC algorithm is nested in the spectrum allocation algorithm to effectively control the conflict problem of F-RAN system agents. Based on the interactive feedback mechanism, the EGFC scheme has the potential to handle multiple targets without using more complex multi-target tracking algorithm.

REFERENCES

Addis, B., Ardagna, D., Panicucci, B., Squillante, M. S., & Zhang, L. (2013). A Hierarchical Approach for the Resource Management of Very Large Cloud Platforms. *IEEE Transactions on Dependable and Secure Computing*, *2*(1), 253–272. doi:10.1109/TDSC.2013.4

Arrow, K. A., Harris, T. E., & Marschak, J. (1951). Optimal inventory policy. *Econometrica*, *19*(3), 250–272. doi:10.2307/1906813

Bao, F., & Chen, I. (2012). Trust management for the internet of things and its application to service composition. *IEEE WoWMoM*, *2012*, 1–6.

Borylo, P., Lason, A., Rzasa, J., Szymanski, A., & Jajszczyk, A. (2016). Energy-aware fog and cloud interplay supported by wide area software defined networking. *IEEE ICC*, *2016*, 1–7.

Botta, A., Donato, W., Persico, V., & Pescape, A. (2014). On the Integration of Cloud Computing and Internet of Things. *IEEE FiCloud*, *2014*, 23–30.

Caton, S., Haas, C., Chard, K., Bubendorfer, K., & Rana, O. F. (2014). A Social Compute Cloud: Allocating and sharing infrastructure resources via social networks. *IEEE Transactions on Services Computing*, *7*(3), 359–372. doi:10.1109/TSC.2014.2303091

Chard, K., Bubendorfer, K., Caton, S., & Rana, O. F. (2012). Social cloud computing: A vision for socially motivated resource sharing. Services Computing. *IEEE Transactions on Service Computing, 5*(4), 551-563.

Chard, K., Caton, S., Rana, O. F., & Bubendorfer, K. (2010). Social cloud: Cloud computing in social networks. In *Cloud Computing* (pp. 99–106). IEEE. doi:10.1109/CLOUD.2010.28

Checko, A., Christiansen, H. L., Ying, Y., Scolari, L., Kardaras, G., Berger, M. S., & Dittmann, L. (2015). Cloud RAN for Mobile Networks - A Technology Overview. *IEEE Communications Surveys and Tutorials*, *17*(1), 405–426. doi:10.1109/COMST.2014.2355255

Chen, X. (2015). Decentralized Computation Offloading Game for Mobile Cloud Computing. *IEEE Transactions on Parallel and Distributed Systems*, *26*(4), 974–983. doi:10.1109/TPDS.2014.2316834

Daniel, K., Jack, L. K., & Richard, H. T. (1986). Fairness and the assumptions of economics. *The Journal of Business*, *59*(4), 285–300.

Dastjerdi, A. V., & Buyya, R. (2016). Fog Computing: Helping the Internet of Things Realize Its Potential. *Computer*, *49*(8), 112–116. doi:10.1109/MC.2016.245

Feng, N., Mau, S. C., & Mandayam, N. B. (2004). Pricing and power control for joint network-centric and user-centric radio resource management. *IEEE Transactions on Communications*, *52*(9), 1547–1557. doi:10.1109/TCOMM.2004.833191

Gajane, P., Urvoy, T., & Clérot, F. (2015). A Relative Exponential Weighing Algorithm for Adversarial Utility-based Dueling Bandits.*Proceedings of the 32nd International Conference on Machine Learning*, (pp. 218-227).

Hemmati, M., Sadati, N., & Nili, M. (2010). Towards a bounded-rationality model of multi-agent social learning in games. *IEEE ISDA*, *2010*, 142–148.

Htikie, Z., Hong, C. S., & Lee, S. W. (2013). The Life Cycle of the Rendezvous Problem of Cognitive Radio Ad Hoc Networks: A Survey. *Journal for Corrosion Science and Engineering*, *7*(2), 81–88.

Hu, Z., Zheng, Z., Wang, T., Song, L., & Li, X. (2016). Game theoretic approaches for wireless proactive caching. *IEEE Communications Magazine*, *54*(8), 37–43. doi:10.1109/MCOM.2016.7537175

Hung, S. C., Hsu, H., Lien, S. Y., & Chen, K. C. (2015). Architecture Harmonization Between Cloud Radio Access Networks and Fog Networks. *IEEE Access*, *3*, 3019–3034. doi:10.1109/ACCESS.2015.2509638

Jang, I., Pyeon, D., Kim, S., & Yoon, H. (2013). A Survey on Communication Protocols for Wireless Sensor Networks. *Journal for Corrosion Science and Engineering*, *7*(4), 231–241.

Jesse, N., Heo, U., & DeRouen, K. Jr. (2002). A Nested Game Approach to Political and Economic Liberalization in Democratizing States: The Case of South Korea. *International Studies Quarterly*, *46*(3), 401–422. doi:10.1111/1468-2478.00239

Jiang, C., Chen, Y., Gao, Y., & Liu, K. J. R. (2013). Indian Buffet Game with non-Bayesian social learning. *IEEE GlobalSIP*, *2013*, 309–312.

Jiang, C., Chen, Y., Gao, Y., & Liu, K. J. R. (2015). Indian Buffet Game With Negative Network Externality and Non-Bayesian Social Learning. *IEEE Transactions on Systems, Man, and Cybernetics. Systems*, *45*(4), 609–623.

Kamijo, Y. (2009). A two-step Shapley value in a cooperative game with a coalition structure. *International Game Theory Review*, *11*(02), 207–214. doi:10.1142/S0219198909002261

Kim, K., Uno, S., & Kim, M. (2010). Adaptive QoS Mechanism for Wireless Mobile Network. *Journal for Corrosion Science and Engineering*, *4*(2), 153–172.

Kim, S. (2010). Dynamic Online Bandwidth Adjustment Scheme Based on Kalai-Smorodinsky Bargaining Solution. *IEICE Transactions on Communications*, *E93-B*(7), 1935–1938. doi:10.1587/transcom.E93.B.1935

Kim, S. (2011). Stackelberg Game-Based Power Control Scheme for Efficiency and Fairness Tradeoff. *IEICE Transactions on Communications*, *E94-B*(8), 2427–2430. doi:10.1587/transcom.E94.B.2427

Kim, S. (2014). *Game Theory Applications in Network Design*. Hershey, PA: IGI Global. doi:10.4018/978-1-4666-6050-2

Kim, S. (2014). Intervenient Stackelberg Game based Bandwidth Allocation Scheme for Hierarchical Wireless Networks. *Transactions on Internet and Information Systems (Seoul)*, *8*(12), 4293–4304.

Kim, S. (2015). Learning based Spectrum Sharing Algorithms by using Coopetition Game Approach. *Wireless Personal Communications*, *82*(3), 1799–1808. doi:10.1007/s11277-015-2314-5

Kim, S. (2016). News-vendor game-based resource allocation scheme for next-generation C-RAN systems. *EURASIP Journal on Wireless Communications and Networking*, (1), 1–11.

Lee, W., Xiang, L., Schober, R., & Wong, V. W. S. (2014). Direct Electricity Trading in Smart Grid: A Coalitional Game Analysis. *IEEE Journal on Selected Areas in Communications*, *32*(7), 1398–1411. doi:10.1109/JSAC.2014.2332112

Li, J., Sun, J., Qian, Y., Shu, F., Xiao, M. & Xiang, W. (2016). A Commercial Video-Caching System for Small-Cell Cellular Networks using Game Theory. *IEEE Access*. (forthcoming)

Liu, Y., Sun, Y., Ryoo, J. W., Rizvi, S., & Vasilakos, A. V. (2015). A Survey of Security and Privacy Challenges in Cloud Computing: Solutions and Future Directions. *Journal for Corrosion Science and Engineering*, *9*(3), 119–133.

Malakooti, B. (2014). *Operations and Production Systems with Multiple Objectives*. New York, NY: John Wiley & Sons.

Pan, M., & Fang, Y. (2008). Bargaining based pairwise cooperative spectrum sensing for Cognitive Radio networks. *IEEE MILCOM*, *2008*, 1–7.

Pang, Y., Xie, S., & Jiang, S. (2011). The application of nested-game theory in the public participation mechanism in the decision-making of large engineering projects. *Systems Engineering Procedia*, *1*, 142–146. doi:10.1016/j.sepro.2011.08.024

Park, H., & van der Schaar, M. (2007). Bargaining Strategies for Networked Multimedia Resource Management. *IEEE Transactions on Signal Processing*, *55*(7), 3496–3511. doi:10.1109/TSP.2007.893755

Park, S., Simeone, O., & Shamai, S. (2016). Joint optimization of cloud and edge processing for fog radio access networks. *IEEE ISIT*, *2016*, 315–319.

Park, Y., & Kim, S. (2015). Bargaining based Smart Grid Pricing Model Demand Side Scheduling Management. *ETRI Journal*, *37*(1), 197–202. doi:10.4218/etrij.15.0114.0007

Peng, M., Yan, S., Zhang, K., & Wang, C. (2016). Fog-computing-based radio access networks: Issues and challenges. *IEEE Network*, *30*(4), 46–53. doi:10.1109/MNET.2016.7513863

Qian, M., Hardjawana, W., Shi, J., & Vucetic, B. (2015). Baseband Processing Units Virtualization for Cloud Radio Access Networks. *IEEE Wireless Communications Letters*, *4*(2), 189–192. doi:10.1109/LWC.2015.2393355

Qiao, H., Rozenblit, J., Szidarovszky, F., & Yang, L. (2006). Multi-Agent Learning Model with Bargaining.*Proceedings of the 2006 Winter Simulation Conference*, (pp. 934-940). doi:10.1109/WSC.2006.323178

Sabyasachi, A.S., De, S. & De, S. (2013). On the Notion of Decoupling in Mobile Cloud Computing. *IEEE HPCC_EUC'2013*, (pp. 450-457).

Sigwele, T., Pillai, P., & Hu, Y. F. (2014). Call Admission Control in Cloud Radio Access Networks. *IEEE FiCloud, 2014*, 31–36.

Singh, D., Tripathi, G., & Jara, A. J. (2014). A survey of Internet-of-Things: Future Vision, Architecture, Challenges and Services. *IEEE World Forum on Internet of Things (WF-IoT' 2014)* (pp. 287-292).

Sinha, K., & Kulkarni, M. (2011). Techniques for fine-grained, multi-site computation offloading. *IEEE CCGRID*, *2011*, 184–194.

Suris, J. E., DaSilva, L. A., Han, Z., & MacKenzie, A. B. (2007). *Cooperative Game Theory for Distributed Spectrum Sharing*. *IEEE, ICC*, 5282–5287.

Tandon, R., & Simeone, O. (2016a). Harnessing cloud and edge synergies: Toward an information theory of fog radio access networks. *IEEE Communications Magazine*, *54*(8), 44–50. doi:10.1109/MCOM.2016.7537176

Tandon, R., & Simeone, O. (2016b). Cloud-aided wireless networks with edge caching: Fundamental latency trade-offs in fog Radio Access Networks. *IEEE ISIT*, *2016*, 2029–2033.

Tsebelis, G. (1988). Nested Games: The Cohesion of French Electoral Coalitions. *British Journal of Political Science*, *18*(2), 145–170. doi:10.1017/S0007123400005044

Vakilinia, S., Qiu, D., & Ali, M. M. (2014). Optimal multi-dimensional dynamic resource allocation in mobile cloud computing. *EURASIP Journal on Wireless Communications and Networking*, 1–14.

Vermesan, O., & Friess, P. (2011). *Internet of Things - Global Technological and Societal Trends*. River Publishers.

Wang, X., & Schulzrinne, H. (2005). Incentive-Compatible Adaptation of Internet Real-Time Multimedia. *IEEE Journal on Selected Areas in Communications*, *23*(2), 417–436. doi:10.1109/JSAC.2004.839399

Wang, Y., Lin, X., & Pedram, M. (2013). A Nested Two Stage Game-Based Optimization Framework in Mobile Cloud Computing System. *IEEE SOSE*, *2013*, 494–502.

Wen, J. (2009). No-centralize Newsvendor Model with Re-distributed Decision-making Power. *IEEE IITA*, *2009*, 237–240.

William, J. S. (2009). *Operations Management* (10th ed.). New York, NY: McGraw-Hill.

Xie, B., Zhou, W., Hao, C., Ai, X., & Song, J. (2010). A Novel Bargaining Based Relay Selection and Power Allocation Scheme for Distributed Cooperative Communication Networks.*IEEE Vehicular Technology Conference (VTC 2010)*, (pp. 1-5). doi:10.1109/VETECF.2010.5594320

Yang, H. (1997a). Sensitivity analysis for the elastic-demand network equilibrium problem with applications. *Transportation Research*, *31*(1), 55–70. doi:10.1016/S0191-2615(96)00015-X

Yang, H., & Bell, M. G. H. (1997b). Traffic restraint, road pricing and network equilibrium. *Transportation Research Part B: Methodological*, *31*(4), 303–314. doi:10.1016/S0191-2615(96)00030-6

Zehavi, E., & Leshem, A. (2009). Alternative Bargaining Solutions for the Interference Channel. *IEEE CAMSAP*, *2009*, 9–12.

Zhang, S. (2014). Influence of relationship strengths to network structures in social network. *IEEE ISCIT*, *2014*, 279–283.

Zhao, Y., & Zhao, H. (2002). Study on negotiation strategy.*International Conference On Power System Technology 2002*, 1335-1338.

Zhu, W. Z., & Lee, C. H. (2014). A New Approach to Web Data Mining Based on Cloud Computing. *Journal for Corrosion Science and Engineering*, *8*(4), 181–186.

KEY TERMS AND DEFINITIONS

Baseband Processing: Baseband processor is used to process the down-converted digital signal to retrieve essential data for the wireless digital system.

Cloud Radio Access Network (C-RAN): C-RAN is a centralized, cloud computing-based architecture for radio access networks that supports 2G, 3G, 4G and future wireless communication standards.

Computation Offloading: The transfer of certain computing tasks to an external platform, such as a cluster, grid, or a cloud.

Kalai-Smorodinsky Bargaining Solution: A solution to the Bargaining problem. It was suggested as an alternative to Nash's bargaining solution. The main difference between the two solutions is that the Nash solution satisfies independence of irrelevant alternatives while the KS solution satisfies monotonicity.

Mobile Cloud IoT: A kind of platform which supports diverse and geographically dispersed devices to work together. Cloud Computing plays a significant role in combining services offered by heterogeneous devices and scaling up to handle large number of users in a reliable manner.

Nash Bargaining Solution: A Pareto efficient solution to a Nash bargaining game. According to Walker, Nash's bargaining solution was shown by John Harsanyi to be the same as Zeuthen's solution of the bargaining problem.

Rubinstein-Stahl Model: Provide a possible solution to the problem that two players are bargaining with the division of the benefit.

Shapley Value: A solution concept in cooperative game theory. To each cooperative game it assigns a unique distribution among the players of a total surplus generated by the coalition of all players. The Shapley value is characterized by a collection of desirable properties.

Social Cloud: A scalable computing model wherein virtualized resources contributed by users are dynamically provisioned amongst a group of friends.

Tragedy of the Commons: An economic theory of a situation within a shared-resource system where individual users acting independently according to their own self-interest behave contrary to the common good of all users by depleting that resource through their collective action.

Chapter 3
IoT System Resource Sharing Mechanisms

ABSTRACT

As the IoT technology continues to grow, it needs to support an increasing range of services. Therefore, IoT networking over which services are provided has become an area of great importance. In particular, the management of IoT resources and the way new technology integrates into the network operator's infrastructure is critical to the success of IoT. The key to supporting a large number of services is IoT system resource. Therefore, all performance guarantees in IoT systems are conditional on currently available resource capacity. In this chapter, we focus our attention on the IoT resource allocation problem. First, an effective bandwidth allocation algorithm for heterogeneous networks is introduced. And then, a new Bitcoin mining protocol with the incentive payment process is explained. To share the computation resource, this Bitcoin protocol adopts the concept of the group bargaining solution by considering a peer-to-peer relationship.

DOI: 10.4018/978-1-5225-1952-2.ch003

PRINCIPAL-AGENT GAME BASED RESOURCE ALLOCATION (PARA) SCHEME

In order to provide more comprehensive network services, a concept of integrated heterogeneous network system was introduced. Until now, lots of researchers have focused on how to efficiently integrate different types of wireless and mobile networks. To exploit the heterogeneous network system operation, an important issue is how to properly manage bandwidth. Recently, S. Kim designed the *Principal-Agent game based Resource Allocation* (PARA) scheme, which is a new bandwidth management algorithm based on the principal-agent game model. Among heterogeneous networks, the PARA scheme has analyzed the asymmetric information situation and developed an effective bandwidth allocation algorithm. Under diverse network condition changes, this principal-agent game approach is essential to provide a suitable tradeoff between conflicting requirements.

Development Motivation

In the past few years, wireless and mobile networks have experienced a great success. However, any single type of existing wireless networks cannot provide all types of services. In order to provide more comprehensive multimedia services, a concept of integrated network system was introduced by combing different types of wireless and mobile networks. In modern times, heterogeneous network systems are used for new business scenarios allowing ubiquitous networking and Internet of Things (IoT) services. To implement these techniques, current trends show that many different wireless networks coexist and cooperate with each other to provide internetworking accesses (Shen, 2008; Xue, 2012).

Based on the interdependence among different networks, the heterogeneous network system is envisaged to be a novel network structure to achieve a high network performance and broad coverage while bringing more flexible and plentiful access options for mobile users. However, it faces great challenges such as resource management problems among multiple networks. Therefore, proper resource management strategies, including bandwidth allocations in each network, are of great importance to exploit the potential network diversity (Shen, 2008; Xue, 2012).

Over the years, a lot of research work has been done to build new resource management schemes and to evaluate them. They have focused on the bandwidth management techniques based on the diversity of the services provided by different networks (Lopez-Benitez, 2011; Xue, 2012). However, they did not provide suitable solutions that ensure the conflict relationship among heterogeneous networks. These existing techniques are one-sided protocols and can not adaptively respond the current network conditions. In the 1970s, the Principal-Agent (PA) game model emerged from the combined disciplines of economics and institutional theory (Eisenhardt, 1989). Traditionally, PA models arise when the incentives between the agent and the principal are not perfectly aligned. As a result, the agent may be tempted to act in his own interest rather the principal's, and conflicts of interest are almost inevitable. Therefore, the PA model has been used to formulate the relationship between an information-advantaged agent and a principal able to issue a contract ultimatum. This game model has come to extend well beyond economics or institutional studies to all contexts of information asymmetry, uncertainty and risk (Eisenhardt, 1989).

The PARA scheme adopts a PA game approach to address the bandwidth allocation problem in heterogeneous networks. Usually, different networks may pursue different interests, and act individually to maximize their profits. This self-organizing feature can add autonomics into network systems and the use of PA game model has expanded for designing the mutually satisfactory bandwidth allocation mechanism. To strike an appropriate system performance, the PARA scheme captures the conflicting relationship among different networks, and incorporates an interactive learning mechanism to find the most profitable solution.

System Architecture and Game Model

Heterogeneous networking architecture is organized based on the cooperation among different networks involved. The PARA scheme takes three different types of wireless networks such as cellular network, WLAN and Wi-Fi networks. WLAN and Wi-Fi networks are connected with the cellular network in the same manner as any other radio access networks (Lohi, 1999; Ning, 2006). Cellular networks consist of a number of cells, and, in each cell, multiple WLAN and Wi-Fi networks exist. As a result, the data traffic from WLAN or Wi-Fi users goes through the cellular network before reaching the Internet or other packet data networks.

Each cell in cellular mobile networks is serviced by a Base Station (BS). In order to support the interworking requirements, BSs are responsible to support integrated authentication, accounting and bandwidth management for traffic services within their corresponding cell areas. Typically, a geographic region in each cell is subdivided into areas; global, local and hotspot areas. Local and hotspot areas are served by WLAN and Wi-Fi networks, respectively. The BS can cover a global area while allocating bandwidth to WLAN and Wi-Fi networks within its own cell. These networks coexist to form a heterogeneous network system and have overlapping areas of coverage to provide services (Lohi, 1999; Ning, 2006). The general architecture of heterogeneous wireless network system is shown in Figure 1.

A key concept of heterogeneous networking is the unification of several heterogeneous networks in each cell. In order to satisfy the unification problem, one of the most important control issues is bandwidth allocation. To efficiently allocate the limited wireless bandwidth to heterogeneous networks, a well-known methodology is the game theory. Usually, a formal game model consists of players, the possible strategies of the players, and utility functions of the strategies. Therefore, to represent a traditional game G, the game model components are given by $G = <$ *Players* ($\mathbb{N}$), *Strategies* ($\mathbb{S}$), *Utility_functions* (U) $>$, where $\mathbb{N}$ is a finite set of players $\{ P_1, P_2, \dots P_n \}$, $\mathbb{S}$ is a non-empty set of the strategies, i.e., $\mathbb{S} = \{ \boldsymbol{S}_1 \dots, \boldsymbol{S}_n \}$, and U is the utility function set ($\{ \boldsymbol{U}_1, \dots, \boldsymbol{U}_n \}$) of players. Each player's satisfaction level can be translated into a quantifiable metric through a utility function. Therefore, a utility function is a measure of the satisfaction experienced by a player; it is a result of a selected strategy. Usually, the utility function maps the user-

Figure 1. Architecture of heterogeneous wireless network system

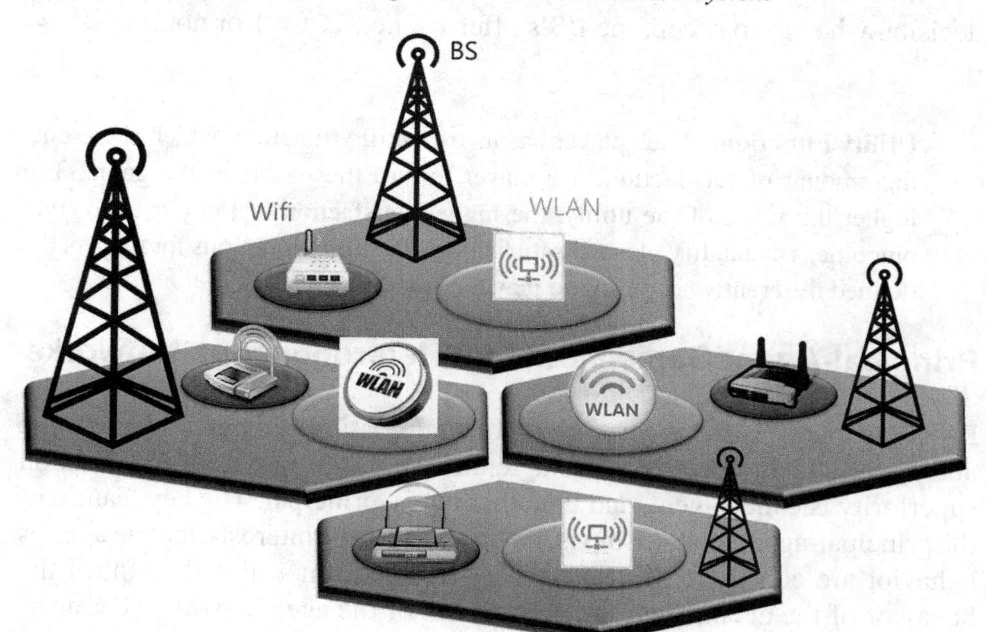

level satisfaction to a real number, which represents the resulting payoff. The goal of each player is to maximize his own payoff by selecting a specific strategy where $\max_{S_i}: U_i(S_i) \rightarrow \Re,\ S_i \in \mathbb{S}$ (Kim, 2014). Based on these notations, the game model for the bandwidth allocation problem can be formulated as follows.

- **Players:** In each cell, there are one BS, and multiple heterogeneous networks such as WLAN and Wi-Fi networks. They are assumed as game players, which are denoted as $\mathbb{N} = \{P_1, P_2, \ldots P_n\}$ where n is the total number of networks within each cell.
- **Strategies:** Each player ($P_{i, 1 \leq i \leq n}$) has strategies. If the player i is the BS, i.e., $i = 1$, a strategy set can be defined as

$$S_1 = \{ (\delta_2, \xi_2), (\delta_3, \xi_3), \ldots, (\delta_n, \xi_n) \}$$

where $\delta_{k, 2 \leq k \leq n}$ and $\xi_{k, 2 \leq k \leq n}$ are the bandwidth allocation policy and cost factor for the kth networks, respectively. If the player k is WLAN or Wi-Fi networks, i.e., $P_k \in \{P_2, \ldots P_n\}$, the S_k is defined as $\{\tau_k, \mathfrak{U}_k\}$ where τ_k is the amount of requested bandwidth from the player k, and $\mathfrak{U}_k = \{0, 1\}$ is the player k 's decision whether to accept the BS's offer δ_k, i.e., $\mathfrak{U}_k = 1$ or not, i.e., $\mathfrak{U}_k = 0$.

- **Utility Functions:** Each player has its own utility function, which represents the amount of satisfaction of a player toward the payoff of the game. The higher the value of the utility, the higher satisfaction of the player for that outcome. To quantify players' satisfaction, the utility functions for players are defined differently according to players' standpoints.

Principal-Agent Game Model for Heterogeneous Networks

PA game theory is the theory of contractual relationship between principal and agents. In the principal-agent relationship, the one with information superiority is called agent, and the other one is principal. The key feature of the principal-agent relationship is that the principal's interests and the agent's behavior are closely related, but the principal can not directly control the behavior of the agent, even the supervision of the agent's work. Therefore, the principal indirectly affects the behavior of the agent only by compensation or wage system (Yuan-xiang, 2013). From the perspective of principals, the principal always hopes to monitor and encourage agents through various ways to let them do their best to work, to complete the responsibility from principals, to create the greatest possible benefit for principals. From the perspective of agents, the agent always wants more rights, get as much return as possible and bear as little responsibility and risk as possible in material and spiritual. Therefore, inevitable conflicts between them result in the principal-agent problem (Liu, 2010).

In the PA game model, the payoff to the principal depends on an action taken by the agent. The principal cannot contract for the action, but can compensate the agent based on some observable signal that is correlated with the action. In general, the principal authorized agent to take corresponding activities and deal with things in a certain range in its own name, including entrusted some decision-making power to an agent to form a power and revenue sharing relationships between the principals and the agents (Liu, 2010). The principal-agent methodology is a powerful tool to model a wider range of real life situations, such as business administration, economics, political science, sociology, psychology, and so on, where informational asymmetry exists. In the PARA scheme, the basic concept of PA game theory is adopted to design a contract mechanism for the bandwidth allocation procedure among heterogeneous networks.

For efficient operations of integrated network infrastructures, the PARA scheme guides selfish heterogeneous networks toward a socially optimal outcome. To satisfy this goal, the wireless bandwidth should be allocated adaptively to each heterogeneous networks considering the context, network capabilities and particular goals. Usually, the role of BSs is to allocate bandwidth to WLAN and Wi-Fi networks within their covering areas, and the role of WLAN and Wi-Fi networks is to adaptively distribute the allocated bandwidth to end users. Therefore, it is suitable assumption that the BS plays the role of a principal and WLAN and Wi-Fi networks become agents. In each cell, one principal, i.e., BS, and n-1 agents, i.e., WLANs and Wi-Fi networks, exist. Let $U_{i,2\leq i\leq n}$ denote the agent i 's utility function like as.

$$U_i = \left[\left(\frac{1-e^{-\beta_j \times \left(b_j^i - mb_j\right)}}{1-e^{-\beta_j \times \left(Mb_j - mb_j\right)}} \right) - \left(\xi_i \times \left(\delta_i\right)^q \times \mathfrak{U}_i \right) - f_i\left(\theta_i\right) \right]$$

s.t.,

$$\theta_i = \sum_{j=1}^{l} b_j^i, f_i(\theta_i) = \left[\sigma \times e^{\frac{\theta_i}{M_i}}\right] \times \mathfrak{U}_i + \rangle_i, \text{ and } \theta_i \leq \delta_i \times \mathfrak{U}(i) \tag{1}$$

where b_j^i is the currently allocated bandwidth of the application service j in the agent i. Mb_j, mb_j are the maximum and minimum bandwidth requests of the application service j, respectively. The value β is chosen based on the average slope of the linear utility function of the request. To reflect the difference of quantitative QoS requirements, the β can quantify the adaptability of an application. ξ_i is the BS's price strategy for the agent i and q is an positive cost factor. $f_i(\cdot)$ is a control overhead function; M_i and $\rangle_i$ are the maximum bandwidth capacity and the basic maintenance cost of the agent i, respectively.

In the formula (1), the first term captures the net income of agent i, and the second term represents the agent i's payment to the BS. It increases with the δ_i; if δ_i is big, the agent i should pay more. The third term means an extra charge for the control overhead and basic maintenance cost. If the agent i actually uses a large amount of bandwidth, it should pay a higher control overhead cost. Finally, each agent's utility is defined as his profit minus the payment and cost. Therefore, agents intelligently decides their strategy τ, i.e., requested bandwidth amount, to maximize their payoff.

For a principal, the payoff is realized based on the amount of bandwidth allocation (δ) and cost factor (ξ) for each agent. Therefore, the utility function (U_1) for the BS can be quantified as follows.

$$U_1 = \sum_{i=2}^{n} \left(\xi_i \times (\delta_i)^q \times \mathfrak{U}_i\right), s.t., \xi_i \in \mathbb{S} = \left[\xi^{\min}, \xi^{\max}\right] \tag{2}$$

where $\xi^{\min}$, $\xi^{\max}$ are the pre-defined minimum and maximum price levels, respectively. Therefore, ξ_i is bounded between the $\xi^{\min}$ and $\xi^{\max}$. The real value of U_1 is determined based on the cost factor (ξ) for agents and agents' decisions $\mathfrak{U}$. During the bandwidth allocation procedure, the principal is a first mover, and chooses the ξ value to ask the payment of agents. Then, agents determine their actions whether to accept the principal's offer or not, considering their expected payoff. Based on the received signals correlated with agents' actions, the principal adjusts dynamically the ξ and δ values.

At each game period, the BS observes agents' τ values in an online manner and decide δ value as follows.

$$\delta_i(t+1) = \frac{\tau_i(t)}{\sum_{k=2}^{n} \tau_k(t)} \times (\mathcal{T}_{\mathcal{B}} - \mathcal{A}_{BS}) \tag{3}$$

where $\delta_i(t+1)$ is the allocated bandwidth for the agent i at time $t+1$, and $\tau_i(t)$ is the agent i's τ value at time t. $\mathcal{T}\_\mathcal{B}$ is the total available bandwidth of BS, and $\mathcal{A}_{BS}$ is the amount of allocated bandwidth for the BS itself. After the decision of δ values, the BS select the price strategy (ξ). To decide adaptively $\xi_{i,2\le i\le n}$ values, the PARA scheme presents a novel dynamic solution using concept from the gradient follower learning approach (Alnwaimi, 2015). At each incremental learning step, the BS obtains a payoff (U_1) as a consequence of ξ value decisions. Based on this information, the BS predicts the future utility at time $t+1$ (U_1^{t+1}), and adaptively adjusts the ξ value. For the agent i, $P_i^{\varepsilon}(t)$ is denoted as the probability weight (propensity) of price strategy ε, i.e., $\varepsilon \in \mathbb{S}$, at time t. The adjusting process for $P_i^{\varepsilon}(\cdot)$ is described in the following.

$$P_i^{\varepsilon}(t+1) = P_i^{\varepsilon}(t) + \left[\alpha_i^t \times \left(\frac{\delta_i(t) - \overline{\delta_i(t)}}{\overline{\delta_i(t)}}\right) \times \frac{\psi(\varepsilon)}{e^{\frac{P_i^{\varepsilon}(t)}{\varphi}} \times \left(1 + e^{\frac{P_i^{\varepsilon}(t)}{\varphi}}\right)^2}\right]$$

s.t.,

$$\overline{\delta_i(t)} = \overline{\delta_i(t-1)} + \left[\lambda_i^{t-1} \times \left(\delta_i(t-1) - \overline{\delta_i(t-1)}\right)\right] \text{ and } \psi(\varepsilon) = \begin{cases} 1, & \text{if } \varepsilon = \xi_i(t) \\ 0, & \text{otherwise} \end{cases} \tag{4}$$

where α and λ are the learning rates in the agent i toward maximizing the utility function, and φ is a positive Boltzmann cooling parameter. Considering (1)-(4), the PARA scheme can set the following maximization problem.

$$\max_{\delta_i,\xi_i,\mathfrak{U}(i)} \mathbb{E}\left[U_1\right], s.t., \begin{cases} (i)\, E\left[U_i\right] > 0 \\ (ii) \max\limits_{\tau_i,\theta_i,\mathfrak{U}_i} \mathbb{E}\left[U_i\right] \end{cases} and\ 2 \le i \le n \tag{5}$$

In the maximization problem (5), the constraint (i) means that agents must receive a positive expected utility if they accept the BS's offer, i.e., *individual rationality*. The agent might be required to make an announcement about what is its request (τ), and it can be free to announce its real demand; it is no longer necessarily truthful. Therefore, agents can misreport their true bandwidth requests. If the PARA scheme considers the *revelation principle* to search for an optimal contract, the principal restricts itself to contracts that the agent finds to report the request truthfully to be optimal. The constraint (ii) insures that the agent's optimal policy is to announce his request (τ) truthfully, i.e., *incentive compatibility*.

The Main Steps of the PARA Scheme

In the PARA scheme, the contract design problem has come to the principal-agent game. To define the optimal contract procedure, the PARA scheme characterizes the conflicting requirements from heterogeneous networks, and strike an appropriate network performance. Based on the real-time interactive feedback process, each network can adapt its behavior and act strategically to achieve a better profit. This approach offers a more realistic model for game players with bounded rationality. The main steps of the PARA scheme are given next.

Step 1: At the initial time, ε strategy propensity $P^{\varepsilon}(\cdot)$ for every agents and τ are equally distributed. This starting guess guarantees that each price strategy enjoys the same selectivity, and the available bandwidth is equally allocated for heterogeneous networks at the beginning of the game.

Step 2: Every game period, game players estimate individually their utility function in an entirely distributed manner. According to (1), agents $P_{i,2\le i\le n}$ adaptively distribute the allocated bandwidth to provide traffic services while maximizing their payoffs.

Step 3: At the end of each game period, the BS in each cell estimate its current payoff, and select the next strategies, i.e., ξ and δ values, according to (3) and (4). The results of these strategy decisions are the input back to the agents' decision process.

Step 4: Based on the received signal from the BS, each agent make a decision, i.e., $\mathfrak{U} = 0$ or 1, in an entirely distributed manner. This decision strongly affects the BS's payoff of next time period; it is estimated using the equation (2).

Step 5: Based on the online feedback learning process, each game players can capture how to adapt their strategies to address the maximization problem in (5).

Step 6: Constantly, game players are self-monitoring the current network situation for the next PA game process; proceeds to Step 2 for the next iteration.

Summary

With the incidence of a variety of heterogeneous networks, bandwidth management problem is an important issue for effective network operations. The PARA scheme is a new bandwidth allocation algorithm based on the game theory. The main goal of the PARA scheme is to maximize network performance while ensuring a better bandwidth efficiency. To satisfy this goal, the PARA scheme adopts the method PA game model, and designs an effective bandwidth allocation algorithm. Considering different perspectives of heterogeneous networks, the PARA scheme intelligently allocates wireless bandwidth using an interactive feedback process.

THE GROUP BARGAINING BASED BITCOIN MINING (GBBM) SCHEME

Bitcoin is a digital cryptocurrency that has generated considerable public interests through a fully decentralized network with an inherently independence from governments or any central authorities. Although its short history has been volatile, the Bitcoin maintains a core group of committed users. Recently, S. Kim looks at the Bitcoin complex structure, and designs the Group Bargaining based Bitcoin Mining (GBBM) scheme while developing a novel incentive payment process. To effectively implement an incentive payment mechanism, the GBBM scheme adopts the concept of the group bargaining solution by considering a peer-to-peer relationship, and it is practically applied to a distributed computation network system. Based on the cooperative game model, the GBBM scheme explores an efficient solution that can maximize Bitcoin users' rewards.

Development Motivation

Conventional monetary systems are based on fiat currencies such as US Dollar, Euro, Chinese Yuan, Korean Won and etc. They are all depend on different governments and uncoupled from precious metals such as gold. Therefore, their value rests on the faith people have in the institutions that create them. Due to this reason, it can cause inevitable and disputable issues about monetary inflation or unfair exchange rate. In modern times, new technologies have created some alternatives for traditional money exchange. For example, online payment techniques have gained a lot of popularity and number of users using online payment services have increased. However, online payment systems like as PayPal or Alipay have always been controversial due to the private information loss (Beikverdi, 2015; Lewenberg, 2015; Luu, 2015)

Nowadays, virtual currencies have made great developmental leaps. Virtual currency is a type of unregulated digital money, which is issued and usually controlled by its developers, and used and accepted among the members of a specific virtual community. Earlier forms of virtual currencies digitally transfer large amounts of money between parties at fast speeds while offering some level of anonymity. However, they were more properly characterized as anonymous payment systems rather than currencies on their own terms. More recent innovation in virtual currency has taken the next step into creating what may be characterized as digital money, serving as both a peer-to-peer payment system as well as a store of value (Bohr, 2014).

In 2009, Satoshi Nakamoto released a new peer-to-peer distributed trustless electronic cash, called Bitcoin. Bitcoins are intrinsically valueless; their worth is decided by those trading in them. Its main achievement is its ability to arrive at a consensus about the valid transaction history in a totally decentralized fashion (Beikverdi, 2015; Luu, 2015). As a decentralized form of currency, Bitcoin offers the opportunity to have nearly anonymous transactions. A transaction is formed when a user digitally signs a hash of the previous transaction where this Bitcoin was last spent. Along with the future owner's public key, Bitcoin incorporates this signature in the transaction. Therefore, any user can verify Bitcoin authenticity by checking the signature chain. When users make a transaction using Bitcoin, a change of ownership over the Bitcoin is sent to a public transaction log. Transactions are bundled into blocks which are linked into chains and broadcasted to the entire network.

To prevent double Bitcoin spending, these all transactions should be verified through cryptographic proof in the peer-to-peer network. This process is called Bitcoin mining (Bohr, 2014; Decker, 2013; Gervais, 2014).

Bitcoin mining is the fundamental concept of Bitcoin operations which must be done in checking all monetary transactions and verifying them. During the Bitcoin mining, Bitcoin uses a hash-based Proof-of-Work (PoW) mechanism to generate blocks. More specifically, a Bitcoin user must find a nonce value that results in a value below a given target when hashed with additional fields. If such a nonce is found, it is appended to a chain as a new block while allowing any user to publicly verify the PoW. Therefore, the chain grows each time when a block is generated. Any user connected to the Bitcoin network can participate in creating a block by finding a valid PoW, and get Bitcoins as a reward. For example, if users successfully generate a block, they're granted a fixed amount of Bitcoins. This is an incentive for them to continuously support the Bitcoin system (Beikverdi, 2015; Bohr, 2014; Luu, 2015). Therefore, the incentive distribution is one of major problems in the Bitcoin mining process.

However, the Bitcoin mining process requires dedicated hardwares and consumes intense amounts of energy (Alqassem, 2014). To make the Bitcoin mining profitable, the idea that miners can improve the mining process has emerged and attracted many users. In recent years, users can form groups, called mining pools, and share Bitcoin rewards among the pool members. To dynamically split the reward into pool members in proportion to their contributed computational power, there is a space for algorithmic improvements (Lewenberg, 2015).

Under widely dynamic Bitcoin network conditions, each individual user can be assumed as intelligent rational decision-makers, and they select a best strategy to maximize their expected payoffs. This situation is well-suited for the game theory. Therefore, the main focus of the GBBM scheme is devoted to develop a novel incentive payment process based on the group bargaining solution. From the viewpoint of group bargaining solutions, each group can be regarded as a bargaining party with a well-defined payoff (Chae, 2004; Chae, 2010; Zhang, 2009). Therefore, this approach is expected in the phenomenon that stresses distributed computation situation. In the GBBM scheme, users are grouped as a mining pool, and multiple mining pools exist in a distributed way. Therefore, dual-level negotiations occur within individual users in each mining pool and across multiple pools, simultaneously. Based on the classical Nash bargaining solution, users and pools bargain with each other at the same time. Under widely different and diversified Bitcoin system situations, this dual bargaining game model is suitable to get a globally desirable performance.

Dual-Level Bargaining Game Model for Bitcoin Networks

For the Bitcoin operation, there is no central authority or fractional reserve system controlling the supply of Bitcoins. Instead, they are generated at a predictable rate such that the eventual total number will be 21 million (Reid, 2011). A process which results in the generation of new Bitcoins, called mining, is performed by individual users for reception of incentives in the form of Bitcoins. This mining process is essentially operations of SHA-256 hashing of values in search of a hash digest smaller than a target value (Dev, 2014). Users keep altering the value of nonce, which results in entirely changing the block's header hash, until a valid hash is found. Therefore, the higher the number of random values a user can generate per second, the higher the chances to meet the target in less time (Alqassem, 2014).

Once this winning hash has been discovered, a new block is added to the block chain and Bitcoin incentives are furnished by the Bitcoin network system to the participating users. Practically, the required average work is exponential to the number of zero bits and can be verified by executing a single hash. Therefore, the Bitcoin mining process is designed to need considerable computational effort, from which the security of the Bitcoin mechanism is derived; this makes mining costly for individual users. Due to this reason, the cooperative mining process is necessary to effectively encourage users to pay this computational cost (Alqassem, 2014; Dev, 2014; Reid, 2011).

In the GBBM scheme, the main contribution is the up-to-date presentation of the cooperative Bitcoin mining algorithm based on the dual-level bargaining game model. The dual-level bargaining model ($\mathbb{G}$) is a tuple

$$\left(\Gamma, \mathbb{N}, \left(\boldsymbol{S}_i\right)_{i\in\mathbb{N}}, \boldsymbol{\mathcal{G}}, \Psi\left(\mathbb{N}\right), \left(U_i\right)_{i\in\mathbb{N}}, \partial, T\right)$$

at each time period t of gameplay.

- Γ is a difficult problem for the Bitcoin generation: The difficulty of solving Γ requires a large amount of computation.
- $\mathbb{N}$ is the finite set of players $\mathbb{N} = \left\{1,\ldots i \ldots n\right\}$ where i represents a Bitcoin user with non-zero computational power.

- $\boldsymbol{S}_i$ is the set of strategies of the player i, $\boldsymbol{S}_i = \left(s_i^1 \ldots s_i^k \ldots s_i^y\right)$. $\boldsymbol{S}_i$ is defined as computation contribution levels to solve the problem Γ. For simplicity, the GBBM scheme assumes that each player's strategy remains constant for all players in one game.
- $\mathcal{G}$ is the set of groups of players $\mathcal{G} = \{G_1,\ldots G_j \ldots G_r\}$. $G_{j,1\le j\le r}$ consists of multiple players with the same computation power contributions. Each group is assumed as an individual mining pool for Bitcoin mining process.
- $\Psi\left(\mathbb{N}\right)$ is a partition function. It splits $\mathbb{N}$ into multiple user-groups;

$$\left(\mathbb{N}\right) \rightarrow \left\{G_j \in \mathcal{G} \mid \mathbb{N} = \bigcup_{1\le j\le r}\left\{G_j\right\}\right\}.$$

- U_i is the payoff received by the player i. Traditionally, the payoff is determined as the obtained outcome minus the cost to obtain that outcome.
- $\boldsymbol{\partial}$ is the vector of Bitcoin incentives for each groups $\boldsymbol{\partial} = \{\partial_1,\ldots\partial_j\ldots\partial_r\}$ where ∂_j is an incentive amount for the G_j. Individual member of each group obtains the same amount of Bitcoin.
- T is a time period. The $\mathbb{G}$ is repeated $t \in T < \infty$ time periods with competitive and cooperative manner.

In the game theory, each player's satisfaction level is translated into a quantifiable metric through a utility function. Therefore, a utility function is a measure of the satisfaction experienced by a player, which is as a result of a selected strategy (Kim, 2014). In the GBBM scheme, the utility function maps the user-level satisfaction to a real number, which represents the resulting payoff. The goal of each user is to maximize his own payoff by adaptively selecting a specific strategy. By considering the reciprocal relationship between obtained benefit and incurred cost, the user i's utility function with the strategy s_i^k $\left(U_i\left(s_i^k\right)\right)$ is defined as follows;

$$U_i\left(s_i^k\right) = \left(\partial_j \big/ \left|G_j\right|\right) \Big/ \left(s_i^k \times \frac{1}{\min\left\{\boldsymbol{S}_i\right\}}\right) \quad s.t., \in G_j, \partial_j \in \boldsymbol{\partial} \; and \; s_i^k \in \boldsymbol{S}_i \tag{6}$$

where $|G_j|$ is the number of group G_j and $\boldsymbol{S}_i$ is the set of strategies of the player i, i.e., computation contribution levels. In this game model, each user independently select his strategy to maximize his payoff; the more a user contributes his computing power, the more he needs Bitcoin incentive. To adaptively select the strategy, the user i has a memory ($\mathbb{M}_i$) to keep record of recent m plays of the game, and looks back at the m most recent plays, i.e.,

$$\mathbb{M}_i = (U_i^{t-m}\left(s_i^{(t-m)}\right),\ U_i^{t-m+1}(s_i^{(t-m+1)}),\ldots,\ U_i^{t-1}(s_i^{(t-1)}))$$

where $U_i^{t-m}\left(s_i^{(t-m)}\right)$ is the U_i at the time $t-m$; $s_i^{(t-m)}$ is the selected the strategy at the time $t-m$ where $s_i^{(t-m)} \in \boldsymbol{S}_i$. At each time t, the selection propensity of s_i^k from $\boldsymbol{S}_i$ $\left(\mathfrak{J}^t\left(s_i^k\right)\right)$ is estimated as follows.

$$\mathfrak{J}^t\left(s_i^k\right) = \begin{cases} \mathfrak{J}^{t-1}\left(s_i^k\right), & \textit{if the } s_i^k \textit{ is not selected at time } t-1 \\ \max\left\{\left(\mathfrak{J}^{t-1}\left(s_i^k\right) + \dfrac{\left\{U_i^{t-1}\left(s_i^k\right) - \left(\dfrac{1}{m} \times \sum_{l=t-m}^{t-1} U_i^l\left(s_i^{(l)}\right)\right)\right\}}{U_i^{t-1}\left(s_i^k\right)}\right), 0\right\}, & \textit{otherwise} \end{cases} \quad (7)$$

Based on the estimated propensity, the s_i^k selection probability from $\boldsymbol{S}_i$ at time t $\left(Pr^t\left(s_i^k\right)\right)$ is defined as;

$$Pr^t\left(s_i^k\right) = \mathfrak{J}^t\left(s_i^k\right) \Big/ \sum_{s_i^z \in \boldsymbol{S}_i} \left\{\mathfrak{J}^t\left(s_i^z\right)\right\} \quad (8)$$

Using (8), each user probabilistically selects his strategy at each time period. Therefore, at each game round, users iteratively adapt their strategies to maximize their payoffs while adaptively responding the current Bitcoin system situations.

In this game model, individual users contribute publicly their computation powers in the Bitcoin network. Based on the contribution levels, users are grouped together. Therefore, multiple groups with different computation contributions are formulated, and these groups become multiple mining pools. In each mining pool, users work together toward to maximize their rewards. In order to model this mechanism, the GBBM scheme divides the Bitcoin mining process into two bargaining classes; intra-pool bargaining and inter-pool bargaining. Conceptually, the GBBM scheme can regard these two bargaining processes are interdependent, but being done simultaneously. This dual-level bargaining approach can reduce the variance of mining while encompassing the network's overall computational power.

Group Bargaining Solution for Bitcoin Systems

The GBBM scheme follows the Group Bargaining Solution (*GBS*) concept, which generalizes the classic Nash bargaining model. *GBS* is the unique bargaining solution that satisfies six axioms - *Symmetry, Pareto optimality, Independence of irrelevant alternatives* and *Invariance with respect to utility transformations, Strong individual rationality, Anonymity*. To formally express these axioms, we need to introduce some terminology. An *agreement point* is any strategy vector $\mathbb{S} = \left(s_1^{(\cdot)},\ldots s_i^{(\cdot)},\ldots s_n^{(\cdot)}\right)$ where $s_i^{(\cdot)} \in \boldsymbol{S}_i$. It is a possible outcome of the bargaining process. A *disagreement point* $\boldsymbol{d} = (d_1,\ldots,d_n)$ is also a strategy profile that is expected to be the result of non-cooperative actions; it means a failure of the bargaining process. The player i has its own utility function $\left(U_i\left(s_i^{(\cdot)}\right)\right)$ and it has also a minimum desired utility value, which is the disagreement point, i.e., zero in this model. Let $\boldsymbol{d} = \left(0_1,\ldots,0_n\right) \in \mathbb{R}^n$ be the disagreement point, and $\mathcal{F}$ be a function $\mathcal{F}\left(\mathbb{S},\boldsymbol{d}\right) \rightarrow \mathbb{R}^n$. If we assume a feasible payoff set is nonempty, convex, closed, and bounded, the *GBS* $= \left(s_1^*\ldots s_i^*\ldots s_n^*\right) = \mathcal{F}\left(\mathbb{S},\boldsymbol{d}\right)$ satisfies the six axioms (Chae, 2004; Chae, 2010; Zhang, 2009).

1. **Strong Individual Rationality:** *GBS* should be better off than the disagreement point. Therefore, no player is worse off than if the agreement fails. Formally, $x_i^* \geq d_i$ for all player i.
2. **Anonymity:** For any permutation $\phi : \mathrm{N} \rightarrow \mathrm{N}$, $\mathcal{F}\left(\phi\left(\mathbb{S}\right),\phi\left(\boldsymbol{d}\right)\right) = \phi\left(\mathcal{F}\left(\mathbb{S},\boldsymbol{d}\right)\right)$.

3. **Pareto Optimality:** *GBS* gives the maximum payoff to the players. Therefore, if there exists a solution $GBS = \left(s_1^* \ldots s_i^* \ldots s_n^*\right)$, it shall be Pareto optimal.
4. **Invariance with Respect to Utility Transformations:** For any linear scale transformation of the function ψ, $\psi\left(\mathcal{F}\left(\mathbb{S},d\right)\right) = \mathcal{F}\left(\psi\left(\mathbb{S}\right),\psi\left(d\right)\right)$. This axiom is also called *Independence of Linear Transformations or scale covariance.*
5. **Independence of Irrelevant Alternatives:** *GBS* should be independent of irrelevant alternatives. If $\mathbf{X}^* \in \mathbb{S}' \subset \mathbb{S}$ and $\mathbf{X}^* = \mathcal{F}\left(\mathbb{S},d\right)$, then $\mathbf{X}^* = \mathcal{F}\left(\mathbb{S}',d\right)$.
6. **Symmetry:** If $\mathbb{S}$ is invariant under all exchanges of users, $\mathcal{F}_i\left(\mathbb{S},d\right) = \mathcal{F}_j\left(\mathbb{S},d\right)$ for all possible players i and j.

The *GBS* extends the traditional Nash solution to a more general class of bargaining problems. Therefore, the *GBS* can illustrate the use of game theoretic tools applied to a real-world distributed Bitcoin environment. In the GBBM scheme, S. Kim designs his *GBS* model based on his assumption. Presentively, it is a Nash solution within each mining pool as well as across pools. Therefore, multiple pools bargain with each other while coordinating individual users' mining activities. To implement the dual-level bargaining process, the developed *GBS* can be formulated as follows. There are r groups $= \{G_1, \ldots G_j \ldots G_r\}$ and each group obtains its corresponding Bitcoin incentive ∂ in ∂. The desirable best solution $\partial = \left\{\partial_1^* \ldots \partial_j^* \ldots \partial_r^*\right\}$ is the *GBS*, and it is obtained through the following maximization problem;

$$\boldsymbol{GBS} = \max_{\partial=\{\partial_1,\ldots\partial_j\ldots\partial_r\}} \prod_{G_j \in \mathcal{G}} \left(\prod_{i\in\mathbb{N}, i\in G_j} \left(\left(\log_{\psi_j}\left(U_i\left(s_i^k\right)+1\right)\right) - d_i\right)^{f_{intra}(i)} \right)^{f_{inter}(G_j)}$$

s.t.,

$$s_i^k \in \boldsymbol{S}_i, \sum_{i\in G_j} f_{intra}\left(i\right) = 1 \; and \; \sum_{G_j\in\mathcal{G}} f_{inter}\left(G_j\right) = 1 \tag{9}$$

where ψ_j is the logarithm base for G_j. $f_{intra}(i)$, $f_{inter}(G_j)$ are the bargaining powers of player i, and the mining group G_j, respectively. In the *GBS*, bargaining powers, i.e., $f_{intra}(i)$ (*or* $f_{inter}(G_j)$), are the relative ability to exert influence over other users (*or* pools). Usually, the bargaining solution is strongly dependent on the bargaining powers. If different bargaining powers are used, the game player with a higher bargaining power obtains a higher payoff than the other players. In this bargaining model, all players in the same mining pool contribute the same computation contribution. Therefore, in the intra-group bargaining, game players, e.g., individual users in the same mining pool, have the same computation contribution level. Therefore, the GBBM scheme sets the bargaining powers equally, i.e., $1/|G|$. However, in the inter-group bargaining, game players, e.g., mining pools, have different bargaining powers. In the GBBM scheme, a received benefit typically follows a model of diminishing returns to scale; user's marginal benefit slowly diminishes with increasing contribution power. Based on this consideration, $f_{intra}(i)$ and $f_{inter}(G_j)$ are defined as follows;

$$\begin{cases} f_{intra}(i) = 1/\left|G_j\right|, \ s.t., i \in G_j \\ f_{inter}(G_j) = \dfrac{\dfrac{1}{1+\left[\omega \times \exp\left(\mathcal{T}(G_j) / \max_{G_r \in \mathcal{G}} \{\mathcal{T}(G_r)\} \right)\right]}}{\sum_{G_r \in \mathcal{G}} \left(\dfrac{1}{1+\left[\omega \times \exp\left(\mathcal{T}(G_r) / \max_{G_r \in \mathcal{G}} \{\mathcal{T}(G_r)\} \right)\right]} \right)} \end{cases} \quad (10)$$

where $\mathcal{T}(G_j)$ is the corresponding strategy of group G_j and ω is the control parameter for group bargaining powers. After Bitcoins have been mined, they are distributed to participating users according to (9). Based on the *GBS*, the obtained Bitcoin is distributed appropriately while ensuring reciprocal fairness.

The Main Steps of the GBBM Scheme

The GBBM scheme formulates a novel group bargaining model for the incentive payment process. Based on a dual-level bargaining game, individual users are willing to compromise their own objective within their corresponding pool. It is a realistic cooperative game model while appropriately reacting to selfish users. The GBBM scheme is described by the following major steps.

Step 1: At the starting time, all control parameters, i.e., n, m, y, $\boldsymbol{S}_i$, ω and q are chosen from the Table 1. Initially, all strategy propensities $\left(\mathfrak{I}\left(s\right)\right)$ are equally distributed.

Step 2: At the current time period, the selection probability $(\Pr\left(s\right))$ is evaluated using (6)-(8). And then, each individual users select dynamically their strategies according to the $\Pr\left(s\right)$.

Step 3: Based on the selected strategies, individual users are grouped into multiple pools.

Step 4: For the intra-pool bargaining, $f_{intra}\left(\cdot\right)$ is decided according to the number of corresponding pool. For the inter-pool bargaining, $f_{inter}\left(\cdot\right)$ is decided according to (10).

Step 5: After bitcoins have been mined, the incentive of each participating user is adaptively decided based the *GBS*; it is calculated while maximizing the formula (9).

Step 6: To reduce the computation complexity, the amount of obtained Bitcoins is assumed to be consist of basic allocation units. Therefore, the value ∂ is multiples of allocation units. This approach is suitable for ultimate practical implementation in real world system operations.

Step 7: At every game round, $U\left(\cdot\right)$, $\mathfrak{I}\left(s\right)$, $\Pr\left(s\right)$, $f_{intra}\left(\cdot\right)$, $f_{inter}\left(\cdot\right)$ and *GBS* are re-evaluated periodically according to (6)-(10).

Step 8: Individual users constantly self-monitor the current Bitcoin network situation in a distributed online manner. For the next game round, go to Step 2.

SUMMARY AND CONCLUSION

Recently, Bitcoin has enjoyed a rapid growth, both in value and in the number of transactions. Its success is mostly due to innovative use of a peer-to-peer network to implement all aspects of a currencies lifecycle, from creation to its transfer between users. However, the current decade has only seen a surge of research activities in single-user mining processes; few research has been done for the cooperative Bitcoin mining mechanism. The GBBM scheme uses cooperative game theory to study how mining pool members are likely to share the computation powers. To gain a fair-efficient Bitcoin mining solution, the GBBM scheme designs a dual-level bargaining game model for the incentive payment process. Based on the desirable features of group bargaining approach, self-regarding users are induced to actively participate in the fair-efficient Bitcoin mining process.

REFERENCES

Alnwaimi, G., Vahid, S., & Moessner, K. (2015). Dynamic Heterogeneous Learning Games for Opportunistic Access in LTE-Based Macro/Femtocell Deployments. *IEEE Transactions on Wireless Communications*, *14*(4), 2294–2308. doi:10.1109/TWC.2014.2384510

Alqassem, I., & Svetinovic, D. (2014). Towards Reference Architecture for Cryptocurrencies: Bitcoin Architectural Analysis. *IEEE iThings, GreenCom and CPSCom*, (pp. 436-443).

Beikverdi, A., & Song, J. S. (2015). Trend of centralization in Bitcoin's distributed network.*IEEE/ACIS International Conference on Computing'*2015, (pp. 1-6). doi:10.1109/SNPD.2015.7176229

Bohr, J., & Bashir, M. (2014). Who Uses Bitcoin? An exploration of the Bitcoin community. *IEEE PST*, *2014*, 94–101.

Chae, S., & Heidhues, P. (2004). A group bargaining solution. *Mathematical Social Sciences*, *48*(1), 37–53. doi:10.1016/j.mathsocsci.2003.11.002

Chae, S., & Hervé, M. (2010). Bargaining among groups: An axiomatic viewpoint. *International Journal of Game Theory*, *39*(1), 71–88. doi:10.1007/s00182-009-0157-6

Decker, C., & Wattenhofer, R. (2013). Information propagation in the Bitcoin network. *IEEE P2P'2013*, (pp. 1-10).

Eisenhardt, K. M. (1989). Agency theory: An assessment and review. *Academy of Management Review*, *14*(1), 57–74.

Gervais, A., Karame, G. O., Čapkun, V., & Čapkun, S. (2014). Is Bitcoin a Decentralized Currency? *IEEE Security and Privacy*, *12*(3), 54–60. doi:10.1109/MSP.2014.49

Kim, S. (2014). *Game Theory Applications in Network Design*. Hershey, PA: IGI Global. doi:10.4018/978-1-4666-6050-2

Lewenberg, Y., Bachrach, Y., Sompolinsky, Y., Zohar, A., & Rosenschein, J. S. (2015). Bitcoin Mining Pools: A Cooperative Game Theoretic Analysis. *Proceedings of the 2015 International Conference on Autonomous Agents and Multiagent Systems*, (pp. 919-927).

Liu, Y. (2010). The Game Model of Logistics Finance Based on the Principal-Agent Theory. *IEEE ICEEE, 2010*, 1–4.

Lohi, M., Weerakoon, D., & Aghvami, A. H. (1999). Trends in multi-layer cellular system design and handover design.*IEEE Wireless Communications and Networking Conference*, (pp. 898 – 902). doi:10.1109/WCNC.1999.796801

Lopez-Benitez, M., & Gozalvez, J. (2011). Common Radio Resource Management Algorithms for Multimedia Heterogeneous Wireless Networks. *IEEE Transactions on Mobile Computing*, *10*(9), 1201–1213. doi:10.1109/TMC.2010.221

Luu, L., Saha, R., Parameshwaran, I., Saxena, P., & Hobor, A. (2015). On Power Splitting Games in Distributed Computation: The Case of Bitcoin Pooled Mining. *IEEE CSF*, *2015*, 397–411.

Ning, G., Zhu, G., Li, Q., & Wu, R. (2006). Dynamic Load Balancing Based on Sojourn Time in Multitier Cellular Systems.*IEEE Vehicular Technology Conference*, 111-116.

Reid, F., & Harrigan, M. (2011). An Analysis of Anonymity in the Bitcoin System. *IEEE PASSAT*, *2011*, 1318–1326.

Shen, W., & Zeng, Q. A. (2008). Resource Management Schemes for Multiple Traffic in Integrated Heterogeneous Wireless and Mobile Networks. *IEEE ICCCN*, *08*, 1–6.

Xue, P., Gong, P., Park, J., Park, D., & Kim, D. (2012). Radio Resource Management with Proportional Rate Constraint in the Heterogeneous Networks. *IEEE Transactions on Wireless Communications*, *11*(3), 1066–1075. doi:10.1109/TWC.2011.102611.110281

Yuan-xiang, J., & Hong-lian, G. (2013). The principal-agent game analysis among accounting firm, enterprise customer and government. *IEEE ICSSSM, 2013*, 623–627.

Zhang, X. (2009). A note on the group bargaining solution. *Mathematical Social Sciences*, *57*(2), 155–160. doi:10.1016/j.mathsocsci.2008.09.001

KEY TERMS AND DEFINITIONS

Bitcoin: A digital asset and a payment system invented by an unidentified programmer, or group of programmers.

Bitcoin Mining: To form a distributed timestamp server as a peer-to-peer network, bitcoin uses a proof-of-work system similar to Adam Back's Hashcash and the internet rather than newspaper or Usenet posts. The work in this system is what is often referred to as bitcoin mining.

Heterogeneous Network System: A network connecting computers and other devices with different operating systems and/or protocols.

Incentive Compatibility: A mechanism is called incentive-compatible if every participant can achieve the best outcome to him/herself just by acting according to his/her true preferences.

Individual Rationality: In fully cooperative games players act efficiently when they form a single coalition, the grand coalition. The focus of the game is to find acceptable distributions of the payoff of the grand coalition. Distributions where a player receives less than it could obtain on its own, without cooperating with anyone else, are unacceptable - a condition known as individual rationality.

Principal-Agent Problem: Occurs when one person or entity, the 'agent', is able to make decisions on behalf of, or that impact, another person or entity, the 'principal'. This dilemma exists in circumstances where the agent is motivated to act in his own best interests, which are contrary to those of the principal, and is an example of moral hazard.

Proof-of-Work (PoW): An economic measure to deter denial of service attacks and other service abuses such as spam on a network by requiring some work from the service requester, usually meaning processing time by a computer.

Revelation Principle: A fundamental principle in mechanism design. It states that if a social choice function can be implemented by an arbitrary mechanism, then the same function can be implemented by an incentive-compatible-direct-mechanism with the same equilibrium outcome.

SHA-256 Hashing: SHA stands for Secure Hash Algorithm. Cryptographic hash functions are mathematical operations run on digital data; by comparing the computed 'hash' to a known and expected hash value, a person can determine the data's integrity. SHA-256 is a novel hash function computed with 32-bit and 64-bit words, respectively.

Virtual Currencies: A type of unregulated, digital money, which is issued and usually controlled by its developers, and used and accepted among the members of a specific virtual community.

Wi-Fi Networks: Wi-Fi is a technology that allows electronic devices to connect to a wireless LAN network, mainly using the 2.4 gigahertz (12 cm) UHF and 5 gigahertz (6 cm) SHF ISM radio bands.

Chapter 4
New Game Paradigm for IoT Systems

ABSTRACT

Game theory is a mathematical language for describing strategic interactions, in which each player's choice affects the payoff of other players. The impact of game theory in psychology has been limited by the lack of cognitive mechanisms underlying game theoretic predictions. Behavioral game, inference game, inspection game and Markov game are recent approaches linking game theory to cognitive science by adding cognitive details, theories of limits on iterated thinking, and statistical theories of how players learn and influence others. These new directions include the effects of game descriptions on choice, strategic heuristics, and mental representation. These ideas will help root game theory more deeply in cognitive science and extend the scope of both enterprises.

DOI: 10.4018/978-1-5225-1952-2.ch004

THE COGNITIVE HIERARCHY THINKING BASED POWER CONTROL (CHTPC) SCHEME

The IoT describes a future world of interconnected physical objects, with several applications in the areas of smart environments. To implement the IoT concept, the research in the areas of power controlled circuits, embedded systems design, network protocols and control theory should be required. With the much advancement in these areas, the realization of IoT is becoming increasingly probable. Recently, S. Kim proposed the *Cognitive Hierarchy Thinking based Power Control* (CHTPC) scheme, which is a novel adaptive power control algorithm for IoT systems. Based on the cognitive hierarchy thinking mechanism, the CHTPC scheme is designed as a new behavioral game model to adaptively control the power level. To effectively solve the power control problem in IoT systems, game theory is well-suited and an effective tool.

Development Motivation

With the rapid development of network technologies over the past decade, IoT becomes an emerging technology for critical services and applications. IoT is a rapidly growing system of physical sensors and connected devices, enabling an advanced information gathering, interpretation and monitoring. In the near future, everything is connected to a common network by an IoT platform while improving human communications and conveniences. Recent research shows more potential applications of IoT in information intensive industrial sectors, and IoT will bring endless opportunities and impact every corner of our world. However, while IoT offers numerous exciting potentials and opportunities, it remains challenging to effectively manage the various heterogeneous components that compose an IoT application in order to achieve seamless integration of the physical world and the virtual one (Singh, 2014; Vermesan, 2011).

Power control has always been recognized as an important issue for multiuser wireless communications. With the appearance of new paradigms such as IoT systems, effective power control algorithms play a critical role in determining overall IoT system performance. According to the adaptively decided power levels, the CHTPC scheme can reduce the interference while effectively improve the system capacity and communication quality. Therefore, the research on power control algorithm in IoT systems is considered an attractive and important topic. However, it is a complex and difficult work under a dynamically changing IoT environment (Ha, 2014).

Usually, there are two different power control algorithms; centralized and distributed power control algorithms. In general, due to heavy control and implementation overheads, centralized control approach is an impractical method. But, a distributed mechanism can transfer the computational burden from a central system to the distributed devices. Therefore, in real world system operations, this distributed power control approach is suitable for ultimate practical implementation. In distributed power control algorithms, individual devices locally make control decisions to maximize their profits. This situation can be seen as a game theory problem (Kim, 2014). In classical game theory, players are assumed to be fully rational, and the rules of the game, payoff functions and rationality of the players are taken as common knowledge. However, in recent decades, there had been many conceptual and empirical critiques toward this justification. Empirical and experimental evidences show that game players are not perfectly rational in many circumstances. These results call for relaxing the strong assumptions of classical game theory about full rationality of players (Camerer, 2003).

In 1997, a game theorist C. Camerer had introduced a new concept of game model, called behavioral game theory, which aimed to predict how game players actually behave by incorporating psychological elements and learning into game theory (Camerer, 1997). Usually, behavioral game theory combines theory and experimental evidence to develop the understanding of strategic behavior needed to analyze economic, political, and social interactions. By using an index of bounded rationality measuring levels of thinking, the behavioral game theory can explain why players behave differently when they are matched together repeatedly (Camerer, 2004a; Camerer, 2004b; Camerer, 2015).

To formulate a power control problem, the CHTPC scheme adopts a non-cooperative behavioral game model. Additionally, the key idea of cognitive hierarchy thinking mechanism is used to improve upon the accuracy of predictions made by standard analytic methods, which can deviate considerably from actual experimental outcomes. Based on the game player's cognitive capability, the CHTPC scheme concentrates on modeling the learning behavior in iterative games, and adjusts the current power level of each IoT device as efficiently as possible.

Game Model for Power Control Algorithm

The CHTPC scheme considers a general distributed IoT system, for example, with multiple source–destination node pairs. Each source node has only one target destination, but generates radio signal interference to all other destination nodes that are not its target destination node. With N source nodes, there are N destinations paired to these sources. In any time slot $t = 1, \ldots, T$, the source node *i*, $i \in \mathcal{N} = \{1, \ldots, N\}$, transmits packets concurrently with other sources. Thus, there are $N-1$ interfering signal packets at each destination node for all t, and there are $N(N-1)$ interfering signals across the IoT system. A general figure of system model is shown in Figure 1.

In the target destination node *j*, the *SINR* over the transmitted packet at time slot t is given as follows (Smith, 2014).

$$\gamma_j(t) = \frac{P_j(t) \times h_j^j(t)}{\sum_{i=1, i \neq j}^{N} \left(P_i(t) \times h_i^j(t)\right) + \sigma_j} \quad (1)$$

where $P_i(t)$ is the transmit power of source node *i* at time t and $h_i^j(t)$ is the average channel gain from the source node *i* to the destination node *j*. σ_j is the power of the background noise at the receiver. The CHTPC scheme follows the assumption in (Kim, 2011; Long, 2007; MacKenzie, 2001; Smith, 2014); device transmitters use variable-rate M-QAM, with a bounded probability of symbol error and trellis coding with a nominal coding gain. Ac-

Figure 1. System model for the CHTPC scheme

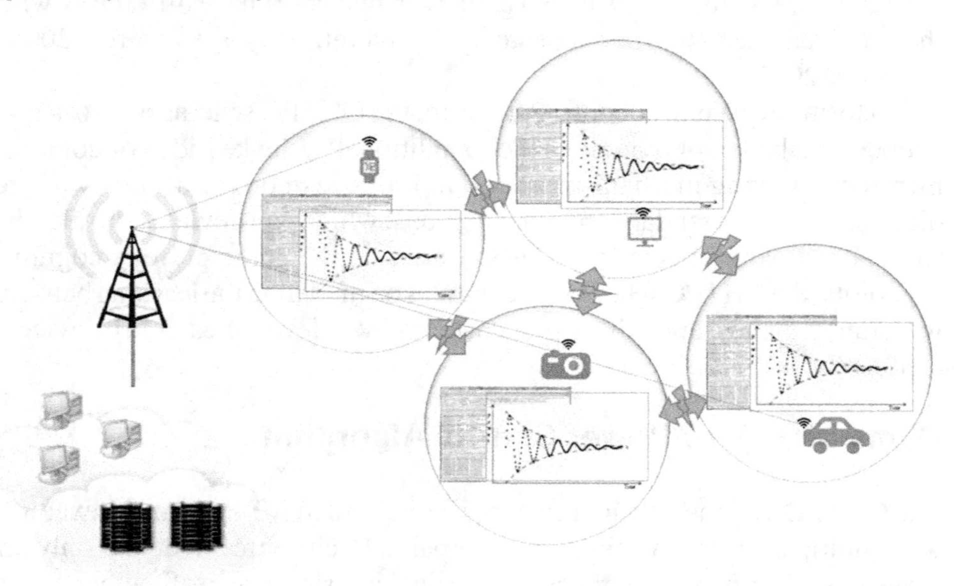

cording to any packet size and data rate, the packet delivery ratio of destination node *j* (PDR_j) can be expressed as a compressed exponential function of the inverse SINR $1/\gamma$.

$$PDR_j\left(P_i, \boldsymbol{P}_{-i}\right) = \exp\left(-\left(\frac{1}{\gamma_j \times \eta}\right)^{\varrho}\right) \quad (2)$$

where γ_j is the node *j*'s SINR. η and ϱ are constant parameters with respect to particular packet sizes and data rates, respectively (Smith, 2014).

The main goal of power control problem is to decide how the co-channel link is shared among different devices while maximizing the total system performance. To effectively solve this problem, the CHTPC scheme adopts the behavioral game model. To design the behavioral game model, game form ($\mathbb{G}$) can be formulated with four parameters: players ($\mathcal{N}$), a strategy set ($\boldsymbol{S}$) for each player, payoffs (U) of the strategies and thinking level (K) of players. Mathematically, $\mathbb{G}$ can be defined as

$$\mathbb{G} = \left\{\mathcal{N}, \left\{\boldsymbol{S}_i\right\}_{i\in\mathcal{N}}, \left\{U_i\right\}_{i\in\mathcal{N}}, K\right\}$$

at each time stage t of gameplay.

- $\mathcal{N}$ is the finite set of players, which are mobile nodes in the IoT systems.
- $\boldsymbol{S}_i$ is the set of strategies with the player *i*. Strategies are power levels (i.e., $P_i \in \boldsymbol{S}_i$) and the range of possible power levels can only take a restricted number of discrete values in the range [$P_i^{\min}, \ldots, P_i^{\max}$] where $P_i^{\max}$ and $P_i^{\min}$ are the pre-defined maximum and minimum power levels, respectively.
- The U_i is the payoff received by the player *i*.
- The K is a thinking level of players.

The behavioral game $\mathbb{G}$ is repeated $t \in T < \infty$ time periods with imperfect information. Therefore, the source node *i*'s decisions are made without the knowledge of opponent players' (i.e., $-\boldsymbol{i}$) decisions. The utility function ($U_i^t\left(P_i\right)$) of the source node *i* at time t is defined as follows; the first term on the right side of the equation (3) represents cost, which is caused by power consumption, and the second term means an outcome, which is received packet amount through wireless communications.

$$U_i^t\left(P_i, \boldsymbol{P}_{-i}\right) = -\left[\kappa_i \times \frac{P_i(t)}{P_i^{\max}}\right] + \left[\theta_i \times \left(PDR_j\left(P_i, \boldsymbol{P}_{-i}\right)\right)^{\xi_i}\right], s.t., \kappa_i, \theta_i, \xi_i > 0 \qquad (3)$$

where κ_i, θ_i and ξ_i are weighting factors for the node i. The bigger values of κ indicate that power saving is more important than the packet delivery ratio, and the relatively smaller κ values to θ and ξ are vice versa. P_i is within the strategy space of player *i*. To maximize individually payoffs, the transmit power should be decided depending on other players' power levels in the system.

Cognitive Hierarchy Thinking Mechanism

Traditional game theory is a mathematical system for analyzing and predicting how game players behave in strategic situations. It assumes that all players form beliefs based on an analysis of what others might do, and choose the best response given those beliefs. However, this assumption is obviously not satisfied under the real world environment; experiments have shown that players do not always act rationally. To redeem this major shortcoming, the behavioral game theory offers a more realistic model for players with bounded rationality. The primary goal of behavioral game theory is to make accurate predictions (Camerer, 2004; Camerer, 2015). To satisfy this goal, the Cognitive Hierarchy (CH) mechanism was developed to provide initial conditions for models of learning while predicting behaviors in non-cooperative games (Camerer, 2004). For the player i, strategy attractions are mapped into probabilities; the selection probability for the l th strategy ($Prob_i^l(t+1)$) for the game round $t+1$ is defined as follows.

$$Prob_i^l\left(t+1\right) = \frac{\exp\left(\lambda \times A_i^l\left(t\right)\right)}{\sum_{k \in \boldsymbol{S}_i} \exp\left(\lambda \times A_i^k\left(t\right)\right)}, s.t., l \in \boldsymbol{S}_i \qquad (4)$$

where λ is the response sensitivity, and $A_i^k(t)$ is the player i's attraction to choose the strategy k at time t. The CHTPC scheme assumes that the players adjust their attractions for each strategy during the game process. If the

λ is infinite, a player gets greedy learning, in which only the action with the highest propensity is taken. If λ approximates zero, all strategies have equal probability. Therefore, the key challenge is to find an adaptive value of λ that achieves a reasonable trade-off (Camerer, 2003). In the CHTPC scheme, λ is decided according to the player's thinking level.

To compute a strategy attraction ($A(\cdot)$), the CHTPC scheme should know the other players' decisions. Reasoning about other players might also be limited, because players are not certain about other players' rationality. In the CH mechanism, the thinking mechanism is modelled by characterizing the number of levels of iterated thinking that subjects do, and their decision rules. If some players are zero-level thinkers, they do not reason strategically at all, and randomize equally over all strategies. Players, who do one-level of thinking, do reason strategically and believe others are all zero-level thinkers. Proceeding inductively, players who are *K*-level thinkers assume that all other players use zero to *K*-1 level thinking. The key issue in CH thinking mechanism is to decide the frequencies ($f(K)$) of K -level thinkers. From a common-sense standpoint, $f(K)/f(K-1)$ should be declining in K ; in general

$$f(K)/f(K-1) \propto 1/K.$$

It turns out to imply that $f(K)$ has a Poisson distribution with mean and standard deviation τ. Therefore, the frequency of level K types is $f(K) = \frac{e^{-\tau} \times \tau^{K}}{K!}$ where τ is an index of the degree of bounded rationality in the population (Camerer, 2003; Camerer, 2004; Camerer, 2015).

Given this consideration, the player *i* using *K*-level thinking computes his attraction ($A_i^l(K \mid t+1)$) for the strategy *l* at the time $t+1$ like as

$$A_i^l(K \mid t+1) = \sum_{h \in \mathbf{S}_{-i}} \left(U_i\left(s_i^l, \mathbf{s}_{-i}^h\right) \times \left[\sum_{c=0}^{K-1} \left(\frac{f(c)}{\sum_{c=0}^{K-1} f(c)} \times Prob_{-i}^h(c \mid t) \right) \right] \right)$$

s.t.,

$$Prob_{-i}^{h}\left(c \mid t\right) = \frac{\exp(\lambda_{c} \times A_{i}^{h}(c \mid t))}{\sum_{e \in S_{-i}} \exp\left(\lambda_{c} \times A_{i}^{e}\left(c \mid t\right)\right)} \text{ and } \lambda_{c} = \frac{1}{1 + \omega \times e^{-\epsilon \times t}} \quad (5)$$

where $Prob_{-i}^{h}(c \mid t)$ is the predicted probability of the lower level thinkers, and λ_{c} is obtained according to the thinking levels (c) of players. h is a strategy for players without the player i ($\boldsymbol{S}_{-i}$). ω and ϵ are the control parameters for responsive sensitivity.

At each stage of behavioral game, players seek to play the best response with the combined effect of all other players' actions (i.e., $\boldsymbol{s}_{-i}^{h}$). According to beliefs about what others will do, players are mutually consistent; that is, each player's belief is consistent with what the other players actually do. Therefore, instead of finding a static equilibrium point, players try to maximize their satisfactions through a cognitive thinking process. All the take together, the CHTPC scheme introduces a new solution concept, called *Mutually Consistent Behavior Equilibrium* (*MCBE*). The *MCBE* is a set of strategies with receiving feedbacks. When a set of strategies has chosen by all players and the change of all players' payoffs are within a pre-defined minimum bound (Λ), this set of strategies constitute the *MCBE*. That is formally defined as follows.

$$MCBE = \left\{\left\{P_{1}^{t} \times \ldots \times P_{i}^{t} \times \ldots \times P_{N}^{t}\right\} \mid \max_{i}\left\{\left(P_{i}^{t} - P_{i}^{t-1}\right) \mid 1 \leq i \leq N\right\} < \Lambda\right\} \quad (6)$$

where N is the total number of players. The *MCBE* is a near-Nash equilibrium. In the *MCBE*, players have no incentives to deviate their beliefs and strategies. Therefore, the *MCBE* can capture the idea that a player will have to take into account the impact of his current strategy on the future strategy of other players

Summary and Conclusion

For the last decades, a new game theory research has relaxed a mutual consistency to predict how players are likely to behave in in one-shot games before they can learn to equilibrate. The CHTPC scheme has looked at a behavioral game model to explain what happens in the player's mind during the course of the creative process. Based on the cognitive hierarchy mechanism, the CHTPC scheme designs a new power control algorithm for IoT systems. This scheme dynamically re-adjusts the current power strategy, and approximates a new solution in an iterative learning methodology. In the CHTPC scheme, strategic thinking, best-response, and mutual consistency are key modeling principles. Therefore, this approach enables a shift from association-based to causation-based thinking, which facilitates the fine-tuning and manifestation of the creative work.

THE SENSOR COMMUNICATION RATE CONTROL (SCRC) SCHEME

In real-life situations, decisions must be made even when limited or uncertain information is available. Therefore, the payoff of an action is not clearly known when the decision is made. Recently, game theory has become a powerful tool for analyzing the interactions between decision makers in many domains. However, the traditional game theory approach assumes that a player belief about the payoff of a strategy taken is accurate. To address this problem, a new kind of game, called an inference game, has been introduced. In this game model, it is studied how degrees of uncertainty of belief about payoffs impact the outcomes of real world games. To approximate an optimal decision, the inference game model can clarify how to better manage ambiguous information. The *Sensor Communication Rate Control* (SCRC) scheme applies the inference game model to the sensor communication paradigm, and confirm that this approach achieves better performance than other existing sensor communication schemes in widely diverse IoT environments.

Development Motivation

The rapid development of Internet of Things (IoT) technology makes it possible to connect various smart objects together through the Internet and to provide more data interoperability methods for application purposes. Recent research shows an increase in the number of potential applications of IoT in information-intensive industrial sectors. In various scenarios, IoT can be realized with the help of sensor communication, which provides ubiquitous networking to connect devices, so that they can communicate with each other to make collaborative decisions with limited, or without any, human intervention. Recently, the sensor communication paradigm has been considered as a new type of communication, empowering full mechanical automation that has the potential to change our life styles (Chen, 2012).

However, enabling sensor communication in IoT is not straightforward. One major issue is how multiple machine-type devices should be connected in dynamic network situations. In addition, to achieve successful sensor communications, Quality-of-Service (QoS) provisioning is another important requirement. For machine devices, some applications require deterministic and hard timing constraints, and disasters occur when these are violated. For other applications, statistical and soft timing constraints are acceptable. Thus, one of the most challenging tasks is how to effectively multiplex massive accesses with enormously diverse QoS characteristics (Lien, 2011). Existing mechanisms do not adaptively tackle this QoS issue when services in IoT are performed. Until now, it is a complex and difficult work in a dynamically changing IoT environment (Giluka, 2014; Yu, 2011).

For IoT multimedia services, decisions that influence QoS are related to the packet rate control for application traffic. Based on real-time feedback, each machine device can adapt its behavior and make QoS decisions strategically to maximize its payoffs (Raazi, 2010). This strategic interaction among machine devices can be formally modeled as a decision-making mechanism. It is regarded as a process that results in the selection of a course of action from several alternatives. However, in real-world IoT operations, control decisions have to be made with only limited information. To address this issue, it is necessary to develop an effective control decision mechanism that works in situations involving uncertainty, which is caused by time pressure, lack of data, unknown factors, randomness outcome of certain attributes, etc. (Xiong, 2012; Xiong, 2014a; Xiong, 2014b).

The fundamental assumption of classical game theory is that the consequence or payoff of a strategy profile is determinate or precise (Dirani, 2006; Park, 2007). However, this assumption seems implausible and unreasonable under the real world environment. In view of realistic situations, game players may not be able to exactly expect their precise payoffs of strategy profiles. Due to limited information, players in real-life games have to make decisions under uncertainty. In canonical opinion, '*uncertainty*' is referred to as a kind of ambiguity that describes situations where decision makers cannot determine a precise probability distribution over the possible consequences of an action (Xiong, 2014). Therefore, in games under uncertainty, the players could only assign a set of possible payoffs, rather than a precise payoff, and have an imprecise probability distribution over this set (Xiong, 2012; Xiong, 2014). To model this situation with indeterminate payoffs, some researchers have tried to apply some original ideas taken from decision theory to game models. However, this kind of work still assumes that the consequences in a game are accurate; it cannot adequately handle the problem concerning uncertain consequences and attitudes of players (Xiong, 2014).

By employing the rule of inferences, the SCRC scheme can allow a player belief concerning the possible payoffs, and determine a preference ordering over actions with respect to expected payoffs. Therefore, this game model can relax the rather stringent assumption of traditional game models. Based on the uncertainty-control game model, the SCRC scheme develops a new packet transmission rate control scheme for sensor communication. In interactive situations involving uncertainty, machine devices in the SCRC scheme can respond to current IoT system conditions for adaptive management. Therefore, they properly select the most adaptable strategy for packet transmissions while ensuring QoS for sensor communication. The distinct feature of the SCRC scheme is a more realistic game-based approach with the limited information.

Inference Game Model and Inference Process

To model strategic interactive situations involving uncertainty, the SCRC scheme develops a new inference game, which is constructed based on the assumption that a player belief regarding the uncertain payoffs. Therefore, an imprecise probability distribution over the set of the possible payoffs is assigned based on the player belief. This means that the game players are not sure about the payoffs of each strategy, but assigns a set of possible payoffs to each strategy profile. To effectively expect the possible payoffs, the SCRC scheme applies some original ideas taken from the Bayesian inference process. For the modeling of uncertainty, this approach has become a key challenge in the real-world decision problems. The inference game model ($\mathbb{G}$) is defined as follows.

Definition 1: An inference game model constitutes a 5-tuple $\mathbb{G} = (\boldsymbol{N}, \boldsymbol{S}, \boldsymbol{\xi}, \mathcal{F}_s, \delta_s)$, where:

1. $\boldsymbol{N}$ is a set of game players,
2. $\boldsymbol{S} = \{s_1, s_2, \ldots, s_n\}$ is a nonempty finite set of all pure strategies of players,
3. $\boldsymbol{\xi} = \{u_1, u_2, \ldots, u_m\}$ is the utility set of all consequent payoffs of each strategy; it is defined as a discrete satisfaction levels of players where $u_{i,1 \le i \le m} \in \mathbb{R}$; there are m level satisfactions,

4. $\mathcal{F}_s = \left\{\mathcal{F}_s\left(u\right) \mid s \in \boldsymbol{S}, u \in \boldsymbol{\xi}\right\}$ is a probability assignment function, which maps the strategy choice s over the strategy set $\boldsymbol{S}$ onto a consequence u over the strategy set ¾ where $\mathcal{F}_s$: $\mathcal{F}_s\left(u\right) \rightarrow [0,1]$ and $\sum_{u \in ¾} \mathcal{F}_s\left(u\right) = 1$,
5. δ_s is the uncertainty degree of consequence, which could be caused by strategy choice s ($0 \leq \delta_s \leq 1$).

During the inference game process, a strategy $s_{k,1\leq k\leq n}$ can cause a consequence $u_{i,1\leq i\leq m}$ that is specified by the mapping probability function $\mathcal{F}_{s_k}$. According to the consequent payoffs, *Expected Payoff Interval* for the strategy s_k ($EPI\left(s_k\right)$) is defined as follows.

$$EPI\left(s_k\right) = \left[U_{\min}\left(s_k\right), U_{\max}\left(s_k\right)\right], s.t., \begin{cases} U_{\min}\left(s_k\right) = \min_{u_i \in \xi}\left\{\mathcal{F}_{s_k}\left(u_i\right) \times u_i\right\} \\ U_{\max}\left(s_k\right) = \max_{u_i \in \xi}\left\{\mathcal{F}_{s_k}\left(u_i\right) \times u_i\right\} \end{cases} \quad (7)$$

Under uncertain situations, $\mathcal{F}_s$ function is essential for the decision makings. In the SCRC scheme, $\mathcal{F}_s$ represents the player belief for outcomes of each strategy. To dynamically adapt the current situation, $\mathcal{F}_s$ is updated as new observations become available. Therefore, it is necessary to adopt a scientific inference method in order to adaptively modify the $\mathcal{F}_s$.

Bayesian inference is a method of statistical inference to provide a logical, quantitative decision. Based on the Bayes' theorem and Bayesian probability rules, Bayesian inference summarizes all uncertainty by a 'posterior' distribution, and gives a 'posterior' belief, which may be used as the basis for inferential decisions. Therefore, the concept of Bayesian inference can be used to provide solutions to predict future values based on historical data. In the SCRC scheme, each player predicts each strategy reliability by using Bayesian inference, and makes a decision for the next round game strategy. During the inference game round, the player has a chance to reconsider the current strategy with incoming information and reacts to maximize the expected payoff. According to the Bayes' theorem and updating rule, the Bayesian inference formula can be expressed as follows (Akkarajitsakul, 2011; Kim, 2012; Pan, 1996).

$$P_t\left(H \mid e\right) = \frac{P_t\left(e \mid H\right) \times P_t\left(H\right)}{P_t\left(e\right)} \tag{8}$$

where $P_t\left(H \mid e\right)$ is the posterior distribution of hypothesis H under the evidence e; t represents t^{th} round of game process. $P_t\left(H\right)$ and $P_t\left(e\right)$ are the prior probability of hypothesis H and evidence e, respectively. In the inference game, the SCRC scheme defines n hypotheses for n payoff levels and m events for m strategies; they are represented as follows.

$$H = \begin{cases} H_1 = u_1 \ payoff \ is \ obtained \\ \vdots \\ H_n = u_n \ payoff \ is \ obtained \end{cases} \quad and \ e = \begin{cases} e_1 = strategy \ s_1 \ is \ selected \\ \vdots \\ e_m = strategy \ s_m \ is \ selected \end{cases} \tag{9}$$

At each strategy, there are n mapping hypotheses about the payoff distribution; these hypotheses mean the satisfaction degrees about the selected specific strategy.

At first, a player doesn't know the payoff propensity of each strategy, but can learn it based on the Bayesian model. In the SCRC scheme, $P_t\left(e_{j,1\leq j\leq m}\right)$ represents the percentage of strategy s_j (i.e., event e_j)'s selection; it is measured by the number of s_j's selection divided by the total number of all strategy selections. $P_t\left(H_{l,1\leq l\leq n}\right)$ represents the occurrence ratio of hypothesis H_l; it is measured by the occurrence number of H_l divided by the total number of all hypotheses occurrences (tn). $P_t(e_j \mid H_l)$ is the event conditional probability, given the H_l selection; it can be computed as follows.

$$P_t\left(e_j \mid H_l\right) = \frac{P_t\left(H_l, e_j\right)}{P_t\left(H_l\right)}, s.t., P_t\left(H_l, e_j\right) = \frac{h\_e_{lj}}{tn} \tag{10}$$

where h_e_{lj} is the number of strategy e_j's selection when the H_l hypothesis occurs. Therefore, after each interaction, the player dynamically updates its corresponding event conditional probability $P_t(e_j \mid H_l)$. Finally, the posterior probability $P_t(H_l \mid e_j)$, which is the occurring probability of hypothesis H_l under the strategy e_j selection circumstance, can be obtained as follows.

$$P_t\left(H_l \mid e_j\right) = \frac{P_t\left(e_j \mid H_l\right) \times P_t\left(H_l\right)}{P_t\left(e_j\right)} \tag{11}$$

Once getting the $P_t\left(H_l \mid e_j\right)$ probability, the player can compute the probability assignment function for the $(t+1)^{\text{th}}$ round strategy selection ($\mathcal{F}_{s_j}^{t+1}\left(u_l\right)$). It is given by

$$\mathcal{F}_{s_j}^{t+1}\left(u_l\right) = \mathcal{F}_{e_j}^{t+1}\left(H_l\right) = P_t\left(H_l \mid e_j\right), \text{ s.t., } s_j \in \boldsymbol{S}, \text{ and } u_l \in \boldsymbol{\xi} \tag{12}$$

where e_j is the event that the strategy s_j is selected and H_l is the hypothesis that the u_l payoff is obtained ($s_j = e_j$ and $u_l = H_l$). According to (12), each player can update his $\mathcal{F}_s\left(u\right)$ values in an iterative feedback manner.

Uncertainty Degree and Inference Equilibrium

To accurately estimate the expected payoff, the SCRC scheme defines the *uncertainty degree* of each strategy. Based on the $EPI\left(s\right)$, the *uncertainty degree* of a specific strategy s_k ($\delta\left(s_k\right)$) is defined as follows.

$$\delta\left(s_k\right) = \frac{U_{\max}\left(s_k\right) - U_{\min}\left(s_k\right)}{\max_{s \in \boldsymbol{S}}\left\{U_{\max}\left(s\right) - U_{\min}\left(s\right)\right\}} \; s.t., 0 \le \delta\left(s_k\right) \le 1 \tag{13}$$

In order to make adaptive decisions, the SCRC scheme needs a preference ordering for strategies. To estimate a strategy preference, the *Expected Payoff* for the strategy s_k ($E\_P\left(s_k\right)$) is defined according to the $EPI\left(s_k\right)$ and *uncertainty degree* ($\delta\left(s_k\right)$).

$$E\_P\left(s_k\right) = U_{\min}\left(s_k\right) + \left[\left(1 - \delta\left(s_k\right)\right) \times \left(U_{\max}\left(s_k\right) - U_{\min}\left(s_k\right)\right)\right] \tag{14}$$

At each strategy selection time, players select their strategy to maximize the $E\_P(s_k)$ (i.e., $\max_{s \in S}\{E\_P(s)\}$). According to the $E\_P(\cdot)$, each player can compute the selection probability for the strategy s_k at the $(t+1)^{th}$ round ($P_{t+1}(s_k)$). It is given by

$$P_{t+1}(s_k) = P_{t+1}(e_k) = \frac{E\_P(s_k)}{\sum_{s_j \in S} E\_P(s_j)} \tag{15}$$

$P_{t+1}(s_k)$ represents the preference of strategy s_k at the $(t+1)^{th}$ game round. Therefore, based on the observation about the strategies' past *expected payoffs*, players can update each strategy preference. With this information, the player can make a better decision for the next strategy selection.

As a solution concept of inference game, the SCRC scheme introduces the *Inference-Equilibrium (IE)*, which is more general than the Nash equilibrium. To define the *IE*, the SCRC scheme introduces the concept of *uncertainty regret* (*UR*); it is a method of comparing alternatives due to Savage (Savage, 1951). In this approach, the SCRC scheme first obtains the *expected payoff* for each strategy and then calculate the *UR* for each alternative. If there are two strategies (i.e., s_k, $s_j \in \boldsymbol{S}$), the *UR* of strategy s_j against the strategy s_k ($\Lambda_{s_j}^{s_k}$) is given by

$$\Lambda_{s_j}^{s_k} = E\_P(s_k) - U_{\min}(s_j) \tag{16}$$

If $\Lambda_{s_j}^{s_k} \le \Lambda_{s_k}^{s_j}$, the strategy s_j is preferred to s_k by players (Xiong, 2014). If the maximum regret of all players is within a pre-defined minimum bound (ε), this strategy profile and the corresponding payoffs constitute the *IE*. Definition 2 mathematically expresses the *IE*.

Definition 2: *Inference-Equilibrium* (*IE*) is a strategy profile that can be obtained by repeating a symmetric game with comparing in obtaining payoffs. The *IE* is a refinement of the Nash equilibrium and it is associated with mixed strategy equilibriums. When a strategy profile has chosen by of all players and the all current strategies' maximum *UR*s are less than ε, this strategy profile and the corresponding payoffs constitute the *IE*. That is formally formulated as

$$\max_{\mathfrak{n}\in N}\left\{\Lambda\left(\mathfrak{n}\right)\mid\Lambda\left(\mathfrak{n}\right)=\max\left\{\Lambda_{s_i}^{s_k}\mid s_i,s_k\in\boldsymbol{S}\right\}\right\}\leq\varepsilon \quad (17)$$

where $\Lambda\left(\mathfrak{n}\right)$ is the maximum *UR* of the player $\mathfrak{n}$. Therefore, the *IE* is a near-Nash equilibrium; the state is that the current strategies regret of all players is within a pre-defined minimum bound (ε). In the SCRC scheme, the existence of *IE* strongly depends on the value of ε. According to the value size, the game model can reach the *IE*. If ε value is very high, most strategy profiles reach the *IE*. If ε value is very low, i.e., a negative value, all possible strategy profiles cannot reach the *IE*.

Utility Function for IoT Systems

In sensor communication, each machine device only sends or receives a small amount of data, and multiple devices can be grouped as clusters for certain management purposes. To manage such massive accesses, QoS requirements such as delay and throughput are needed for different types of sensor communication services. The SCRC scheme follows the assumption in (Yu, 2011) to implement the sensor services; a *p*-persistence CSMA/CA system with *L* classes of devices - class 1 (*or L*) corresponds to the highest (*or* lowest) priority service. The system totally has $\sum_{i=1}^{L} n_i$ devices, where n_i represents the number of the *i*-th class devices. The traffic activities of the *i*-th class devices follow the Poisson process with mean arrival rate λ_i and departure rate μ_i. In principle, the setting of parameter *p* in *p*-persistent CSMA/CA is equivalently to tuning the size of backoff window in CSMA/CA. If the channel is idle, the device will transmit a packet with probability p_i when new time slot commences. Otherwise, it will wait until the channel is idle. By varying the parameter p_i for the *i*-th class devices, differential QoS provisioning could be easily achieved. For simplicity, the SCRC scheme supposes an M/D/1 queuing model with no packet collisions. Therefore, the average output packet rate of the queuing system is equal to the input rate λ_i. Let T_s^i denote the transmission time of a class *i* device, and the time fraction of that device occupies the channel is given by ($\lambda_i \times T_s^i$). Let ϱ_i represent the probability that the channel is idle for a device of class *i* in a given slot (Yu, 2011).

$$\varrho_i = 1 - \sum_{j=1, j \neq i}^{L} \left(n_j \times \lambda_j \times T_s^j\right) - \left(\left(n_i - 1\right) \times \lambda_i \times T_s^i\right) \tag{18}$$

For the device of class i, the transmission probability in an arbitrary slot is represented by ($\varrho_i \times p_i$). Following the M/D/1 queuing model, the average service rate of the i-th class device (μ_i) and the queuing delay (W_Q^i) is given by

$$\mu_i = \frac{\varrho_i \times p_i}{T_s^i} \text{ and } W_Q^i = \frac{\rho_i}{2 \times \mu_i \times \left(1 - \rho_i\right)} \tag{19}$$

Consequently, the total delay of the i-th class device (d_i) is given by

$$d_i = W_Q^i + W_S^i = \frac{2 - \rho_i}{2 \times \mu_i \times \left(1 - \rho_i\right)}, s.t., W_S^i = \frac{1}{\mu_i} \tag{20}$$

where the average service time W_S^i is the inverse of the average service rate (Yu, 2011). Let $\rho_i = \frac{\lambda_i}{\mu_i}$ denote the utilization coefficient of the i-th class of devices. Finally, d_i can be obtained as follows.

$$d_i = \frac{2 - \left(\frac{\lambda_i}{\mu_i}\right)}{2 \times \mu_i \times \left(1 - \frac{\lambda_i}{\mu_i}\right)} = \frac{\left(2 \times \mu_i\right) - \lambda_i}{2 \times \left(\mu_i^2 - \lambda_i \times \mu_i\right)}, s.t., \rho_i = \frac{\lambda_i}{\mu_i} \ll 1 \tag{21}$$

In the SCRC scheme, the rate control process is formulated as a n-player inference game. The packet transmission contending devices in IoT are game players and each device has its own class. As a game player, a class i device selects a rate λ_i in its strategy space $\mathcal{R}_i \in \left[\lambda_{\min}^i, \lambda_{\max}^i\right]$ to send packets, and then it will gain a payoff according to the selected strategy. In the SCRC scheme, $\mathcal{R}_i \cong \boldsymbol{S}$ and $\lambda_{\min}^i, \lambda_{\max}^i$ are s_1, s_n, respectively. Therefore, available strategies in $\mathcal{R}$ is defined as discrete and multiple packet transmission rates.

Utility functions quantitatively describe the players' degree of satisfaction with respect to its action in the game. In this model, the utility function is defined by

$$u_i = \lambda_i - \left(\omega_i \times d_i\right), s.t., \omega_i > 0 \tag{22}$$

where a tunable parameter ω_i indicates the relative importance weight (delay versus transmission rate) of the i-th class devices. To allow the differential QoS provisioning, the higher priority applications have a larger ω value and the lower priority applications have a smaller ω value. By combining (18)-(22), the SCRC scheme obtains the explicit expression of the utility function as follows.

$$u_i = \lambda_i - \left(\omega_i \times \frac{\left(2 \times \mu_i\right) - \lambda_i}{2 \times \left(\mu_i^2 - \lambda_i \mu_i\right)}\right) = \lambda_i - \left(\omega_i \times \frac{2 \times \left(\frac{\xi_i \times p_i}{T_s^i}\right) - \lambda_i}{2\left(\left(\frac{\xi_i \times p_i}{T_s^i}\right)^2 - \lambda_i \times \left(\frac{\xi_i \times p_i}{T_s^i}\right)\right)}\right)$$
$$= \lambda_i + \left(\frac{\left\{w_i \times T_s^i \times \left(\left[T_s^i \times \lambda_i\right] - \left[2 \times p_i \times \xi_i\right]\right)\right\}}{\left(2 \times p_i \times \left[p_i \times \xi_i + T_s^i \times \lambda_i\right] \times \xi_i\right)}\right) \tag{23}$$

From (23), the CHTPC scheme knows the utility function is actually a function of transmission rate λ_i for all services. Finally, the payoff of the ith class devices depends not only on its own strategy but also the other players' strategies. Therefore, it is represented by the $u_i\left(\lambda_i, \boldsymbol{\lambda}_{-i}\right)$ where »$_{-i}$ is the set of strategies of all devices without the device i. In the inference game, all the devices aim to maximize their payoffs (i.e., maximizing the transmission rate while minimizing the access delay). Let $u_i^*\left(\lambda_i, \boldsymbol{\lambda}_{-i}\right)$ be the maximum payoff for the device i, and it is used as an index to classify received payoffs into four categories; *bad*, *average*, *good* and *excellent* satisfaction levels. In the SCRC scheme, there is a one-to-one relationship between each category and hypothesis. Therefore, each hypothesis represents the 'level of satisfaction'. According to (23), each categories can be mapped into each hypothesis (H) as follows.

$$H = \begin{cases} H_1 = \textit{excellent payoff is gained,} & \textit{if } u_i^C(\lambda_i, \boldsymbol{\lambda}_{-i}) > \Omega_1 \times u_i^*(\lambda_i, \boldsymbol{\lambda}_{-i}) \\ H_2 = \textit{good payoff is gained,} & \textit{if } \Omega_1 \times u_i^*(\lambda_i, \boldsymbol{\lambda}_{-i}) \geq u_i^C(\lambda_i, \boldsymbol{\lambda}_{-i}) > \Omega_2 \times u_i^*(\lambda_i, \boldsymbol{\lambda}_{-i}) \\ H_3 = \textit{average payoff is gained,} & \textit{if } \Omega_2 \times u_i^*(\lambda_i, \boldsymbol{\lambda}_{-i}) \geq u_i^C(\lambda_i, \boldsymbol{\lambda}_{-i}) > \Omega_3 \times u_i^*(\lambda_i, \boldsymbol{\lambda}_{-i}) \\ H_4 = \textit{bad payoff is gained,} & \textit{if } \Omega_3 \times u_i^*(\lambda_i, \boldsymbol{\lambda}_{-i}) \geq u_i^C(\lambda_i, \lambda_{-i}) \end{cases} \quad (24)$$

where $u_i^C(\lambda_i, \boldsymbol{\lambda}_{-i})$ is the currently obtained payoff and $\Omega_{i,1\leq i\leq 3}$ is a threshold parameter for the event classification.

Summary

The SCRC scheme has investigated an uncertainty-control game and packet transmission rate control algorithm for IoT systems. With the uncertainty about payoffs, the SCRC scheme develops a new inference game model, and then reveal how ambiguity degrees of belief about consequences impact the outcomes of a game. On the basis of the inference game, the SCRC scheme is designed as a new sensor communication algorithm. The issue addressed in the SCRC scheme will be increasingly important with the proliferation of IoT applications. In the future, it is interesting to extend this inference game model to various decision-making processes.

THE COGNITIVE RADIO BASED SPECTRUM CONTROL (CRSC) SCHEME

The rapid development of IoT systems has escalated the demand of spectrum resources. However, most of the spectrum resources have been allocated to primary users and the residual ones may not meet the need. Cognitive radio technique has been proposed as a solution to address the problem of spectrum scarcity. The main cognitive radio functionality is accomplished in a cycle composed of spectrum sensing and spectrum sharing. The *Cognitive Radio based Spectrum Control* (CRSC) scheme is a new cognitive radio spectrum control scheme for IoT systems. In order to ensure cooperative spectrum sensing and sharing, a new concept is added, called reciprocal fairness, and used a game theoretical tool in the CRSC scheme.

Development Motivation

The IoT represents a world-wide network structure of heterogeneous cyber-physical objects such as sensors, actuators, smart devices, smart objects, RFID, and embedded computers. They will be seamlessly embedded into the global information network while becoming active participants in business, logistics, information and social processes (Fortino, 2014). To efficiently manage these devices, control mechanisms are required for multiple aspects e.g., privacy of user data, communication resilience, routing with resource-constraints, and energy efficiency. Most of all, the most important challenge with this multi-faceted problem space is an efficient spectrum management. To maximize the IoT system performance, spectrum control strategy becomes a key factor and has been actively studied in modern times (Jang, 2013; Htike, 2013).

In spite of the emergence of high network infrastructures, wireless spectrum is still an extremely valuable and scarce resource. Especially, the rapid development of IoT along with ubiquitous computing has escalated the demand of spectrum resources. However, most of the spectrum resources have been allocated to primary users and the allocated spectrum resources are not efficiently utilized. To make full use of the idle spectrum, the Cognitive Radio (CR) technique has been proposed (Lee, 2015; Mukherjee, 2013). The main feature of CR technique is the capability to share the wireless spectrum with licensed users in an opportunistic manner. In the CR system, two types of users are considered; licensed users (i.e. primary users) and unlicensed users (i.e. secondary users). A primary user (PU) has exclusive access to designated spectrum while a secondary user (SU) is allowed to temporally occupy the idle spectrum which the PU does not use. If PUs come back to use their designated spectrum, SUs should release the momentary-using spectrum and try to find other idle spectrum band (Kim, 2014; Mukherjee, 2013).

For the realization of CR techniques, spectrum sensing is the ground work, and has gotten more and more attentions. Recently, the cooperative spectrum sensing method has been presented as an effective sensing way to improve the performance of sensing in CR systems (Mukherjee, 2013). Cooperative spectrum sensing occurs when a group of SUs voluntarily contribute to sensing and share their local sensing information to get a better picture of the spectrum usage. However, sensing work of SUs consumes a certain amount of energy and time. Therefore, selfish but rational SUs do not serve a common sensing work and tend to be a 'free-rider' to reduce their control overhead. In this instance, SUs face the risk of having no one sense the spectrum. Due to this reason, the key issue with the cooperative sensing method is how to make a selfish SU collaborate with others (Lee, 2015). This situation can be seen as a game theory problem.

In 1962, M. Dresher introduced the fundamental notion of Inspection Game (Dresher, 1962), and treated in greater generality by M. Maschler (Maschler, 1966), in the context of checking possible treaty violations in arms control agreements. As an applied field of game theory, an inspection game is a two-person (i.e., inspector and inspectee) zero-sum multistage game where an inspector verifies that another party, called inspectee, adheres to certain legal rules. Usually, an inspectee has a potential interest in violating these rules, and an inspector attempts to induce compliance with desired behaviors. A mathematical analysis can help in designing an optimal inspection scheme, where it must be assumed that an illegal action is executed strategically. This situation is defined as an inspection game model (Avenhaus, 2004).

Fairness is another prominent issue for the cooperative spectrum sensing. If the concept of fairness is not considered explicitly at the design stage of spectrum sensing algorithms, different SUs' payoffs can result in very unfair. But, despite the concerns of fairness issue in CR systems, not much work has been done in this direction. Usually, fairness is one type of social preference. Therefore, to define the concept of fairness, there is considerable disagreement on what exactly is meant. The CRSC scheme considers a concept of reciprocal fairness, the belief that one should receive what one is due based on one's actions. As a social construct, reciprocal fairness means that in response to friendly actions, people are frequently much nicer and much more cooperative than predicted by the self-interest model; conversely, in response to hostile actions they are frequently much more nasty and even brutal (Kamas, 2012; Rabin, 1993). Based on this assumption, the CRSC scheme investigates the reciprocal fairness issue to stimulate cooperation among SUs, and build up a reciprocal utility function, which can lead to an effective cooperation.

The advantage of using the inspection game is that there is no need to assume infinitely-repeated game process in the CR environment since the CRSC scheme can punish short run players as well. Usually, dynamic game models assume implicitly that the interaction of players lasts for infinite many times, which is generally not true for real world network environments. Also, the CRSC scheme considers the overhead to decide how much effort to perform actions. From the inspector's point of view, the rate of inspection has to be adapted to strike a balance between the audit costs and the improved total system performance. From the inspectees' point of view, the rate of cooperative sensing is decided to maximize their own payoff. In conjunction with the reciprocal fairness, the CRSC scheme can provide an opportunity to balance the system performance.

Inspection Game Model for CR Networks

In the real world, there are many situations where authorities have preferences over individuals' choices. In these situations, a fundamental problem for authorities is how to induce compliance with desired behavior when individuals have incentives to deviate from such behavior (Nosenzoa, 2014). Game theory enables the modelling of such an interdependent decision situation where a game model consists of players with non-aligned interests. By means of game theory, we are able to analyze the misbehavior/auditing scenario under specified circumstances. This situation can be mapped to a specific class of game theory models, called inspection game. In the inspection game model, an inspector controls the correct behavior of inspectees, and administers a sanction if a misbehavior is detected during the inspection (Gianini, 2013).

Under dynamically changing CR environments, the CRSC scheme considers that $\mathcal{N} = \{1,,,n\}$ SUs, $\mathcal{M} = \{1,,,m\}$ PUs and one Cognitive Base Station (CBS). SUs, PUs and CBS have heterogeneous characteristics. The allocated radio spectrum for each PU is exclusively used by the corresponding PU, but PUs may not be active all the time. SUs can opportunistically utilize the spectrum when it is available through spectrum sensing. Multiple cognitive SUs form an ad-hoc cluster to jointly access PU spectrums. When a PU's spectrum is idle, this idle spectrum band is temporarily allocated to SUs in corresponding ad-hoc cluster (Lee, 2015; Mukherjee, 2013).

For the effective spectrum sensing, SUs sense the PU spectrum by using the centralized cooperative sensing method, where each SU senses the spectrum and sends the sensing information to the CBS. By using appropriate data fusion rules, the CBS makes a judgment whether or not the spectrum is idle. Based on the sensing data from multiple SUs, the decision about the spectrum idleness is made collaboratively. Therefore, the cooperative sensing approach through multiple SUs cooperation can increase the sensing accuracy. However, SUs are intuitively reluctant to cooperatively sense and would prefer to free-ride without contributing anything. It is the well-known free-ride problem. To avoid this situation, inspection game model is helpful (Kim, 2013; Kim, 2014).

To design the inspection game model for cooperative sensing algorithm, the CRSC scheme investigates an inspector-inspectees scenario where an inspector (i.e., CBS) can either inspect or not and inspectees (i.e., SUs) can either sensing or non-sensing. In the CRSC scheme, the inspection game form ($\mathbb{G}$) can be formulated with four parameters: players ($\mathbb{N}$), a strategy set ($\boldsymbol{S}$) for each player, payoffs (U) of the strategies and time period (T) of game process. Mathematically, $\mathbb{G}$ can be defined as

$$\mathbb{G} = \left\{ \mathbb{N}, \left\{ \boldsymbol{S}_i \right\}_{i \in \mathcal{N}}, \left\{ U_i \right\}_{i \in \mathcal{N}}, T \right\}$$

at each time stage t of gameplay.

- $\mathbb{N}$ is the finite set of players, which are one inspector (i.e., CBS) and multiple inspectees (i.e., SUs) in the ad-hoc cluster of CR network systems; $\mathbb{N} = \mathcal{N} \cup \left\{ CBS \right\}$.
- $\boldsymbol{S}_i$ is the set of strategies with the player *i*. If the player *i* is an inspector (CBS), $\boldsymbol{S}_i = [\, \mathcal{I}_R^{\min}, \mathcal{I}_R^{\max} \,]$ is the inspection rate ($\mathcal{I}_i^R \in \boldsymbol{S}_i$). If the player *i* is an inspectee ($i \in \mathcal{N}$), $\boldsymbol{S}_i = [\, \mathcal{S}_R^{\min}, \mathcal{S}_R^{\max} \,]$ is the sensing rate ($\mathcal{S}_i^R \in \boldsymbol{S}_i$).
- The U_i is the payoff received by the player *i*. If the player *i* is an inspector, U_i is defined as the total system performance minus the inspection cost. If the player *i* is an inspectee, U_i is defined as individual SU's payoff.
- The T is a time period. The inspection game $\mathbb{G}$ is repeated $t \in T < \infty$ time periods with imperfect information.

Inspection Game Based Cooperative Sensing Algorithm

In the classical game theory, the assumption is that game players are perfectly rational, and individually pursuit the maximization of their own personal payoffs. However, this traditional hypothesis has been challenged since 1980s. Recently, more and more behavioral economists and psychologists have accumulated much evidence to flesh out the idea that game players relatively think that psychological payoff is equivalently important than material payoff to reach the self-fulfilling status. In other word, players prefer reciprocal altruism and have fair thinking in games (He, 2010; Rabin, 1993; Tavoni, 2009). Therefore, the concept of reciprocal fairness should be considered for the new model of game theory.

To develop a new cooperative sensing algorithm, the CRSC scheme constructs a utility functions for SUs by relying on the influence of reciprocal interaction. Instead of assuming that utility is either monotonically increasing or decreasing based on its outcome, the SU's utility function is decided not only depending his own received payoff, but also considering reciprocal fairness. When matched with other SUs, the SU *i*'s utility function $\left(U_{i,i\in\mathcal{N}}\left(x_i,\boldsymbol{x}_{-i},s_i\right)\right)$ is defined as follows.

$$U_i\left(x_i,\boldsymbol{x}_{-i},s_i\right)=\mathcal{B}_i\left(x_i,\boldsymbol{x}_{-i}\right)-\lambda\times\mathcal{C}_i\left(s_i\right), s.t., i\in\mathcal{N} \text{ and } s_i\in\boldsymbol{S}_i=\left[\mathcal{S}_R^{min},\mathcal{S}_R^{max}\right]$$

$$\begin{cases}\mathcal{B}_i\left(x_i,\boldsymbol{x}_{-i}\right)=\left(x_i-\left[\frac{1}{n-1}\times\left(\frac{1}{\alpha_i^j}\times\sum_{j\neq i}\max\left\{x_j-x_i,0\right\}\right)\right]\right)/\pi_i^{\max}\\ \mathcal{C}_i\left(s_i\right)=s_i/\max\left\{\boldsymbol{S}_i\right\}\end{cases} \quad (25)$$

where $\mathcal{B}_i(\cdot)$ and $\mathcal{C}_i(\cdot)$ are the received benefit and sensing cost functions, respectively. The parameter λ controls the relative weights given to benefit and cost. The CRSC scheme develops the $\mathcal{B}_i(\cdot)$ function by including a reciprocal fairness issue to provide cooperative stimulus. In $\mathcal{B}_i(\cdot)$, x_i and x_j are the SU *i* and SU *j*'s material outcomes, respectively ($i\neq j$ and $i,j\in\mathcal{N}$). Therefore, the CRSC scheme can specify the received benefit in such a way that the SU *i* suffers if he gets less than other SUs, but he is relatively indifferent with respect to other SUs' benefit if he is better off himself. The first term in the right-hand side in $\mathcal{B}_i(\cdot)$ represents a material outcome. The second term in the right-hand side in $\mathcal{B}_i(\cdot)$ is the psychological outcome loss from disadvantageous inequality (Tavoni, 2009). Therefore, the alternative utility model is linear in both inequality aversion and in the outcomes. $\pi_i^{\max}$ is the maximum spectrum amount for the SU i. Therefore, $\mathcal{B}_i(\cdot)$ and $\mathcal{C}_i(\cdot)$ are normalized and lie in the interval $[0,1]$. α_i^j is the SU *i*'s '*kindness*' to the SU *j*. It represents how kind SU i is being to SU j. Many scholars capture the norm of '*kindness*' as a social capital factor; it has important influence on individual's behavior (He, 2010; Rabin, 1993). The CRSC scheme assumes that SUs have a shared notion of kindness with reciprocal fairness. To incorporating reciprocal fairness into the utility function, α_i^j is given by

$$\alpha_i^j = \left(\frac{\pi_i^c}{\pi_i^{\max} - \pi_i^{\min}}\right) \Bigg/ \left(\frac{\pi_j^c}{\pi_j^{\max} - \pi_j^{\min}}\right) \quad s.t., i \neq j \in \mathcal{N} \tag{26}$$

where $\pi_i^{\max}, \pi_i^{\min}$ $(or\, \pi_j^{\max}, \pi_j^{\min})$ are the SU *i*'s (*or* the SU *j*'s) maximum and minimum spectrum amounts, respectively. $\pi_i^c\, (or\, \pi_j^c)$ is the SU *i*'s (*or* the SU *j*'s) currently allocated spectrum amount. If $\alpha_i^j > 1$, the SU i has a generosity to the SU j in direct proportion to the quantity α_i^j value. Under rationality and reciprocity, the main goal of each SU is to maximize his own payoff by selecting a specific strategy.

For the effective spectrum sharing, the CBS adaptively allocate the idle spectrum band to SUs. The CRSC scheme adopts the notion of *relative utilitarianism* based on the social choice theories and welfare economics (Dhillon, 1999). Therefore, the outcome of CBS is defined according to the *Relative Utilitarian Bargaining Solution* (*RUBS*); it is a natural calibration mechanism, and adaptively allocates the available spectrum to maximize the SU's utilitarian sum as follows

$$RUBS\left(x_{i, i \in \mathcal{N}}\right) = \max_{x_i} \sum_{i \in \mathcal{N}} \left(\frac{\mathfrak{Y}_i - \pi_i^{\min}}{\pi_i^{\max} - \pi_i^{\min}}\right)^{\rho_i^t}$$

s.t.,

$$\mathfrak{Y}_i = x_i - \left(\frac{1}{n-1} \times \left(\frac{1}{\alpha_i^j} \times \sum_{j \neq i} \max\left\{x_j - x_i, 0\right\}\right)\right) \tag{27}$$

where ρ_i^t is the SU *i*'s bargaining power at the time period *t*. It is the relative ability to exert influence over other SUs. Usually, the bargaining solution is strongly dependent on the bargaining power. If different bargaining powers are used, the SU with a higher bargaining power obtains a more spectrum resource than other SUs. In the CRSC scheme, the ρ is used to induce SUs' cooperative behaviors.

In this game model, the ρ is defined as an index of SU's cooperation behavior, which is determined appropriately based on SU's sensing actions.

Due to the dynamically changing CR network environment, SU behaviors can vary spatially and temporally, and ρ value of each SU is dynamically changeable. To effectively adapt this fluctuation, the CBS needs to monitor each SU's behavior constantly. For the implementation of time-dependent monitoring procedure, the CRSC scheme partitions the time-axis into equal intervals of length *unit_time*, and each SU's action is examined periodically at every *unit_time*. After the tth time period, ρ_i^t is defined as the current ratio of the number of SU's cooperative sensing (α_i^t) to the total number of inspections ($\alpha_t^i + \beta_t^i$).

$$\rho_i^t = \frac{\alpha_i^t}{\alpha_i^t + \beta_t^i}, s.t. \begin{cases} \alpha_i^t = \alpha_i^{t-1} + 1, \beta_i^t = \beta_i^{t-1}, & \textit{if the SU i cooperatively sensed} \\ \alpha_i^t = \alpha_i^{t-1}, \beta_i^t = \beta_i^{t-1} + 1, & \textit{otherwise} \end{cases} \tag{28}$$

According to (28), ρ_i^t is a general average function over the whole span of spectrum sensing records. For a long-term period evaluation, the α_i^t and β_i^t would be accumulated and are growing into a very large value. In such case, the recent information will be hard to impact on the overall rating of bargaining power. To solve this problem, attenuation window was introduced (Xiang, 2013). By considering the more current records, the CRSC scheme can calculate the ρ_i^t while fading away the out-of-time records. Based on the attenuation window, the α_i^t and β_i^t values is obtained as below.

$$\alpha_i^t = \sum_{\delta=k}^{n} e^{-\left(\frac{n+m-t(\delta)}{c}\right)} \text{ and } \beta_i^t = \sum_{\delta=j}^{m} e^{-\left(\frac{n+m-t(\delta)}{c}\right)}, s.t., e^{-\left(\frac{n+m-t(\delta)}{c}\right)} \in [0,1] \tag{29}$$

where *e* is the Euler's constant, and *c* is the coefficient to adjust the speed of decreasing in the results of α_i^t and β_i^t. The *n* and *m* are the numbers of cooperative sensing and free-riding, respectively. The *k* and *j* are the most out-of-time period for cooperative sensing and free-riding, respectively. $t(\delta)$ is the time point when δ occurs (Xiang, 2013). While time is growing bigger and bigger, the value of ($n+m-t(\delta)$) will become smaller and smaller, and finally $e^{-\left(\frac{n+m-t(\delta)}{c}\right)}$ has a strong impact on the recent information. Moreover, the bigger value of coefficient *c*, the slower in speed of decreasing slopes of the value in $e^{-\left(\frac{n+m-t(\delta)}{c}\right)}$. In such way, attenuation window can emphasize the most up-to-time records and fade away the out-of-time records by the speed controlled by the coefficient c. Therefore, the attenuation window method can effectively update SUs' ρ values according to a sequence of SU action observations. In the CRSC scheme, the CBS maintains the bargaining power information of all SUs by using the attenuation window.

According to (27),(28) and (29), the utility function of CBS $\left(U\left(x_{i,i\in\mathcal{N}}, s_i\right)\right)$ is defined as the same manner like as the utility function of SUs. The parameter ζ controls the relative weights given to the total system performance and inspection cost.

$$U_k\left(x_{i,i\in\mathcal{N}}, s_k\right) = \mathcal{T}_k\left(x_{i,i\in\mathcal{N}}\right) - \zeta \times \mathcal{C}_k\left(s_k\right)$$

s.t.,

$$k \in \mathbb{N}, k \notin \mathcal{N}, s_k \in \boldsymbol{S}_k = \left[\mathcal{I}_R^{\min}, \mathcal{I}_R^{\max}\right]$$

$$\begin{cases} \mathcal{T}_k\left(x_{i,i\in\mathcal{N}}\right) = RUBS\left(x_{i,i\in\mathcal{N}}\right) \Big/ \max\left\{RUBS\left(x_{i,i\in\mathcal{N}}\right)\right\} \\ \mathcal{C}_k\left(s_k\right) = s_k \Big/ \max\left\{\boldsymbol{S}_k\right\} \end{cases} \tag{30}$$

Finally, the CRSC scheme formulates a new inspection game model for CBS and SUs. Usually, non-cooperative sensing rate of each individual SU strongly depends from inspection rate. This overall interdependent interaction made by the CBS and SUs is modelled as an inspection game. In the inspection game model, the CBS chooses the inspection rate, and each individual SU chooses the cooperative sensing rate; they have conflicting preferences. From the viewpoint of CBS, the main goal is to maximize the CR system performance while considering the *relative utilitarianism* and inspection overhead. From the viewpoint of SUs, the main goal is to maximize their own payoffs, which are defined by capturing the key features of reciprocal fairness.

Aspiration Equilibrium for CR Networks

Nash equilibrium, named after John Nash, is a well-know and classical solution concept in non-cooperative games. It is a set of strategies such that no player has incentive to unilaterally change his action. Since the introduction of the Nash equilibrium concept, it is considered as a de-facto standard solution in game theory. However, recently, substantial evidence has accumulated on the limited ability of the Nash equilibrium concept. When the Nash equilibrium is applied to non-competitive environments, misleading predictions can be naturally occurring. This realization has led to several attempts aimed at finding another reasonable solution concept which performs better empirically (Kim, 2014; Tavoni, 2009).

Associated with the Nash equilibrium, the CRSC scheme develops a new solution concept, called *Aspiration Equilibrium* (*AE*). *AE* is a psychological reference point while balancing aspiration. Based on learning direction theory, *AE* is applicable to the repeated choice in learning situations where game players receive feedbacks not only about the payoff for the choice taken, but also for the payoffs connected to alternative actions. The CRSC scheme assumes that *AE* is a set of strategies chosen by all players, and formally defined as follows.

$$AE = \left\{ \left\{ s_i \right\}_{i \in \mathbb{N}} \mid \min_i \left\{ \left(U_i^t - A\_U_i^t \right) \mid i \in \mathbb{N} \right\} \geq 0 \right\} \tag{31}$$

where U_i^t and $A\_U_i^t$ are the utility and aspiration level of player i at the time period t. Aspiration level (A_U) is the expected degree of future achievement, and initialized individually for each game player. From a commonsense stand point, the CRSC scheme assumes that A_U decreases as time passes. Based on a monotonous time decreasing function ($\delta(\cdot)$), A_U_i is dynamically adjusted as follows.

$$A\_U_i^{t+1} = \delta(t) \times A\_U_i^t, s.t., \delta(t) = 1 - \left(\frac{e^{\xi(t)} - e^{-\xi(t)}}{e^{\xi(t)} + e^{-\xi(t)}} \right) \text{and } \xi(t) = e^{t/\theta} \tag{32}$$

where t is the time period and θ is a control constant. Usually, traditional Nash equilibrium concept doesn't take *game dynamics* into account. However, the *AE* solution can capture the idea to incorporate game dynamics into inspection game model. In the *AE*, players are satisfied based on the impact of his current strategy, and no incentives to deviate their selected strategies.

To approximate the *AE*, the CRSC scheme designs a new learning-based inspection game process. At each stage of inspection game, the CBS and SUs attempt to select the best strategy with the combined feedback information. Individual SUs try to get the strategy using a utility function of the parameters such as material and psychological outcomes. The CBS also attempts to find the best strategy of inspection using a utility function of the parameters such as system outcomes and cost of individual inspections. In the game model, the CBS and SUs can take account of previous strategies to update their beliefs about what is the best-response strategy in the future. If a strategy change can bring a higher payoff, players have a tendency to move in the direction of that successful change, and vice versa. Therefore, the CBS and SUs dynamically tune their strategies based on the payoff history. The SU i's strategy at the time period $t+1$ (s_i^{t+1}) is defined like as

$$\begin{cases} s_i^{t+1} = \Lambda\left[s_i^t + \left|\Delta_s\right|\right], & if\ \Delta_d > 0 \\ s_i^{t+1} = \Lambda\left[s_i^t - \left|\Delta_s\right|\right], & if\ \Delta_d \leq 0 \end{cases} \text{ and } \Lambda\left[\mathcal{K}\right] = \begin{cases} \Lambda\left[\mathcal{K}\right] = \mathcal{S}_R^{\min}, & if\ \mathcal{K} < \mathcal{S}_R^{\min} \\ \Lambda\left[\mathcal{K}\right] = \mathcal{K}, & if\ \mathcal{K} \in \boldsymbol{S}_i \\ \Lambda\left[\mathcal{K}\right] = \mathcal{S}_R^{\max}, & if\ \mathcal{K} > \mathcal{S}_R^{\max} \end{cases}$$

s.t.,

$$i \in \mathbb{N}, \Delta_s = \left(U_i^t - U_i^{t-1}\right) \Big/ U_i^{t-1} \text{ and } \Delta_d = \left(s_i^t - s_i^{t-1}\right) \Big/ \Delta_s \qquad (33)$$

where U_i^{t-1} and s_i^{t-1} are the player i's utility and strategy at the time period $t-1$, respectively. $\left|\Delta_s\right|$ represents the absolute value of Δ_s. For the CBS's strategy at the time period $t+1$, $\mathcal{S}$ is replaced by $\mathcal{I}$ in the equation (33).

Summary

The main issue with the cooperative spectrum sensing is how to make a selfish SU collaborate with others. This situation can be modeled by capturing the key features of inspection game. For this reason, the CRSC scheme designs a novel cooperative sensing algorithm based on the inspection game model. The basic idea of CRSC scheme comes from the realization that SUs' behaviors in non-cooperative situations reflect considerations of fairness. At the same time, the CBS also attempts to balance the system performance while tuning his inspection rate. Therefore, the main novelty of CRSC scheme lies in the study for the interactive feedback approach under realistic CR based IoT scenarios. In dynamic IoT system environments, the CRSC scheme's approach can more evocatively reflect the effective nature of learning attitude for cooperation during non-cooperative repeated game.

R-LEARNING BASED QoS CONTROL (RQSC) SCHEME

The R-learning based QoS Control (RQSC) scheme is formulated as a new QoS management scheme based on the IoT system power control algorithm. Using the emerging and largely unexplored concept of the R-learning algorithm and docitive paradigm, system agents can teach other agents how to adjust their power levels while reducing computation complexity and speeding up the learning process. Therefore, the RQSC scheme can provide the ability to practically respond to current IoT system conditions and suitable for real wireless communication operations.

Development Motivation

The IoT is regarded as a new technology and economic wave in the global information industry after the Internet (Li, 2014; Zhang, 2012). For effective IoT management, more QoS attributes must be considered, including information accuracy, coverage of IoT, required network resources, and energy consumption. Therefore, with regard to IoT services, QoS has been a popular research issue. Typically, different IoT applications have different QoS requirements. The requirements of QoS in an IoT system must be guaranteed while implementing effective resource allocation and scheduling (Duan, 2011; Lin 2012).

In current methods, in order to provide QoS to applications, real-time QoS schedulers are introduced in the IoT structure (Li, 2014). A QoS scheduler is an IoT system component designed to control the IoT system resources for various application services. In each local area, QoS schedulers assign the system resources to contending agents based on a set of criteria, namely, transmitter power, transmission rate, and QoS constraint (Lin 2012). During IoT system operations, these QoS schedulers aim to maximize system utilization while providing QoS requirements to classes of applications that have very tight requirements such as bit rate and delay. However, it is challenging to practically balance between network availability and QoS ensuring (Li, 2014; Lin 2012).

In the RQSC scheme, a new game model, called Team Game (TG) is developed. In the TG model, QoS schedulers are assumed as game players. All game players organize a team, and actions of all players are coordinated to ensure team cooperation by considering a combination of individual payoff as a team payoff. The main concept of TG is to extend the well-known Markov Decision Problem (MDP) to the multi-player case. Traditionally, game theory assumes that players have perfect information about the game, enabling them to calculate all of the possible consequences of each strategy selection (Kim 2014). However, in real world situations, a player must make decisions based on less-than perfect information. If a player does not have total information about the game, then it follows that the player's reasoning must be heuristic. Therefore, to maximize IoT performance, the way in which agents learn network situations and make the best decisions by predicting the influence of others' possible decisions is an important research issue in the field of IoT networking (Jiang, 2014; Martin, 2012; Vrancx, 2008).

In recent years, many learning algorithms have been developed in an attempt to maximize system performance in non-deterministic settings (Vrancx, 2008). Generally, learning algorithms guarantee that collective behavior converges to a coarse equilibrium status. In order to make control decisions in real-time, QoS schedulers should be able to learn from dynamic system environments and adapt to the current network condition. Therefore, QoS schedulers using learning techniques acquire information from the environment, build knowledge, and ultimately improve their performance (Li, 2014; Martin, 2012; Vrancx, 2008).

In 1993, A. Schwartz introduced an average-reward reinforcement learning algorithm, called R-learning algorithm (Schwartz, 1993). Like as Q-learning, R-learning algorithm uses the action value representation. In addition, it also needs to learn the estimate of the average reward. Therefore, R-learning algorithm is performed by the two time-scale learning process. In contrast to the value-iteration based learning algorithms, the decision-learning approach in the R-learning algorithm allows an agent to directly learn the stationary randomized policy, and directly updates the probabilities of actions based on the utility feedback (Mahadevan, 1996).

Recently, there has been increasing interest in research of various R-learning algorithms. However, in this field, many problems still remain open. Even though the R-learning algorithm has received strong attentions, designing a R-learning algorithm for real-world problems is still difficult. There are many complicated restrictions, which are often self-contradictory and variable with the dynamic real-world IoT environment. Docitive paradigm is an emerging technology to overcome the current limitations of R-learning algorithm (Giupponi, 2010). Based not only on the cooperation to learn, but on the process of knowledge transfer, this paradigm can significantly speed up the learning process and increase precision. The docitive paradigm can provide a timely solution based on knowledge sharing in a cooperative fashion with other players in the IoT system, which allows game players to develop new capacities for selecting appropriate actions (Blasco, 2010; Giupponi, 2010).

Markov Decision Process and R-Learning Algorithm

MDP is a mathematical framework for modeling decision making in situations, and is useful for optimization problems solved via reinforcement learning. Based on inputs to a dynamic system, MDP probabilistically determines a successor state, and continues for a finite or infinite number of stages (Galindo-Serrano, 2010; Vrancx, 2008; Wal, 1997). Traditionally, MDP is defined as a tuple $\langle \boldsymbol{S}, \mathbb{A}, T, C(s(t), a(t)), \mathcal{T} \rangle$ where $\boldsymbol{S} = \{s_1, s_2 \ldots s_n\}$ is the set of all possible states and $\mathbb{A} = \{a_1, a_2 \ldots a_m\}$ is a finite set of admissible actions. $T = \{0, 1, \ldots, t, t+1, \ldots\}$ is a time, which is represented by a sequence of time steps. Let $s(t), a(t)$ be the state and action at time step t, respectively where $s(t) \in \boldsymbol{S}$ and $a(t) \in \mathbb{A}$. $C(s(t), a(t))$ is a cost at time t when the action $a(t)$ in state $s(t)$ occurs, and $C(s(t), a(t)) \rightarrow \mathcal{R}$. $\mathcal{T}$ is a state transition function $\mathcal{T} : \boldsymbol{S} \times \mathbb{A} \rightarrow \Delta(\boldsymbol{S})$ where $\Delta(\boldsymbol{S})$ is the set of discrete probability distributions over the set $\boldsymbol{S}$. With the current state and action, the $\mathcal{T}$ function probabilistically specifies the next state of the environment (Galindo-Serrano, 2010; Li, 2014; Vrancx, 2008; Wal, 1997).

The objective of the MDP is to find a policy that minimizes the cost of each state $s(t)$. If Δ is the complete decision policy, the optimal cost value of state $s(t)$ can be written as

$$V^*(s(t)) = \min_{a(t)} \left[C(s(0), a(0)) + \sum_{t=1}^{\infty} \left(\beta^t \times C(s(t), a(t)) \right) \right] \quad (34)$$

where $0 \leq \beta^t < 1$ is a discount factor for the future state. Based on the principle of Bellman's optimality, the equation (34) can be re-written by a recursive equation like as (Galindo-Serrano, 2010; Li, 2014):

$$V^*(s) = \min_a \left[C(s,a) + \beta \times \sum_{s' \in S} \{ P(s' \mid s,a) \times V^*(s') \} \right], s.t., a \in \mathbb{A} \tag{35}$$

where s' represents all possible next states of $\boldsymbol{S}$ and $P(s'|s,a)$ is the state transition probability from the state s to the state s'. If we define Δ as the complete decision policy, we can specify the optimal policy as follows;

$$\Delta^*(s) = \arg\min_a \left[C(s,a) + \beta \times \sum_{s \in S} \{ P(s' \mid s,a) \times V^*(s') \} \right] \tag{36}$$

To solve the equation (36), reinforcement learning algorithms are common way. There exist several reinforcement learning algorithms. The RQSC scheme adopts a novel average-reward reinforcement learning algorithm, call R-learning. Like other reinforcement learning algorithms, R-learning algorithm uses the action value representation (Mahadevan, 1996; Schwartz, 1993). The action value $\mathfrak{R}^{\Delta}(s,a)$ represents the average adjusted value of doing an action a in state s once, and then following policy Δ subsequently (Mahadevan, 1996). That is,

$$\mathfrak{R}^{\Delta}(s,a) = C(s,a) - \eta^{\Delta} + \sum_{s' \in S} \{ P(s' \mid s,a) \times V^*(s') \} \tag{37}$$

s.t.,

$$V^*(s') = \min_{a \in \mathbb{A}} \mathfrak{R}^{\Delta}(s',a)$$

where η^{Δ} is the average reward of policy Δ. In the R-leaning algorithm, system agents learn how to operate in the environment based on the effect of action and reward signal. In order to expediting the interaction between agent and environment efficiently, R-leaning algorithm is performed dynamically by a two-time-scale learning process (Mahadevan, 1996; Schwartz,

1993). However, conventional R-learning algorithm still has the limitation that the learning process of reinforcement learning system will be very slow.

QoS-Aware Service Management in the Team Game Model

To develop a novel QoS-aware service management algorithm, the RQSC scheme assumes and simplifies the real-world situation for practical implementations.

1. Power strategies in each QoS schedulers are quantified. Usually, practicality is decided based on the computation complexity. Therefore, power levels should be simplified.
2. The RQSC scheme assumes four QoS schedulers' system in a local area like as a conference room or small office part. Therefore, the RQSC scheme has been developed as a 4-player game model.
3. Pre-defined minimum bound (ε) is introduced for stable system status. It can be defined according to a subjective point of view for the system stability.
4. Heterogeneous traffic services are categorized into two classes according to the required QoS: class I (real-time) traffic services and class II (non real-time) traffic services. Class I data services are highly delay sensitive, and strict deadlines are applied. However, some flexible data services are called as class II traffic services; they are rather tolerant of delays.
5. If the model is applied in the situation with hundreds or thousands of QoS schedulers in a huge area, each QoS schedulers must be grouped and clustered in a distributed manner. Using a locally distributed approach, the RQSC scheme is applied iteratively in each cluster.

The RQSC scheme considers a new power control mechanism for IoT systems. Under the multi-QoS schedulers' environment, the RQSC scheme formulates the multiple decision-making process as a new TG model based on the multi-agent R-learning approach. Mathematically, the TG model ($\mathbb{T}$) can be defined as

$$\mathbb{T} = \left\{ \boldsymbol{S}, N, \boldsymbol{A}_{i,i\in N}, T, U_{i,i\in N}, T \right\}$$

at each time period $t \in T$ of gameplay.

- S is the set of all possible system states, which is the combinations of power levels of QoS schedulers.
- N is the set of all QoS schedulers.
- $A_i = \{\mathcal{A}_1^i, \mathcal{A}_2^i, \ldots, \mathcal{A}_m^i\}$ is the collection of power levels for the QoS scheduler $i \in N$ where m is the number of possible power levels. $\mathcal{A}_1^i$ and $\mathcal{A}_m^i$ are the pre-defined minimum and maximum power levels.
- U_i is the payoff received by the QoS scheduler $i \in N$.
- T is the state transition function and gives the probability to change the next state.

Usually, traditional solution concept of game theory is obtained with impractical assumptions:

1. Fully rational players,
2. Complete information,
3. Static game-model setting.

These assumptions only hold in the theoretical and idealistic analysis. In the real-world IoT operations, it is impossible to transform the dynamic setting to static simulation setup. It means that the traditional solution with fully rational players is technically impossible to be obtained in the real-worlds. To practically design the TG model, the RQSC scheme develops a new solution concept, called *Stable Equilibrium* (*SE*). Based on the R-learning and docitive paradigm, *SE* is applicable to the repeated choice in learning situations. For the TG, the RQSC scheme assumes that *SE* is a discrete set probability distributions over the available strategies chosen by all players.

Power Control Algorithm Based on the R-Learning Algorithm

The RQSC scheme focuses on how to tackle the QoS control problem based on the R-learning algorithm approach. According to the self-adaptability, QoS schedulers in the RQSC scheme can update the strategy based on the observations while responding to current IoT system conditions. Usually, the main interest of each QoS scheduler is to maximize the amount of transmitted data with the lower power consumption. However, there is a fundamental tradeoff. To capture this conflicting relationship, a utility function (U) is defined by considering the ratio of the throughput to transmit power. In the TG game model, the utility function for the ith QoS scheduler (U_i) is defined as follows:

$$U_i\left(\mathcal{A}_j^i, \mathbb{A}_{-i}\right) = {T_i\left(\mathbb{A}\right)} \Big/ {\mathcal{A}_j^i}, \text{ s.t., } \mathcal{A}_j^i \in \boldsymbol{A}_i \text{ , } \mathbb{A} = \left\{\Pi_{i \in N} \mathcal{A}_i \mid \mathcal{A}_i \in \left[\mathcal{A}_1^i, \mathcal{A}_m^i\right]\right\} \tag{38}$$

where $\mathbb{A}_{-i}$ is the transmit power vector without the $\mathcal{A}_i$, and $T_i\left(\mathbb{A}\right)$ is the throughput of ith scheduler. Usually, throughput is defined as the number of information bits that are transmitted without error per unit-time. In wireless communications, it can be achieved with the SINR (γ_i) in the effective range (Kim, 2014). Therefore, the throughput of the ith scheduler ($T_i\left(\mathbb{A}\right)$) can be expressed as:

$$T_i\left(\mathbb{A}\right) = W \times \log_2\left(1 + {\gamma_i\left(\mathbb{A}\right)} \Big/ {\Omega}\right) \tag{39}$$

where W is the assigned channel bandwidth, and Ω ($\Omega \geq 1$) is the gap between uncoded M-QAM and the capacity, minus the coding gain (Kim, 2014). Finally, the ith scheduler's utility is defined as follows:

$$\max_{\mathcal{A}_j^i \in \boldsymbol{A}_i} U_i = \max_{\mathcal{A}_j^i \in \boldsymbol{A}_i} \left\{ \left[W \times \log_2\left(1 + {\gamma_i\left(\mathbb{A}\right)} \Big/ {\Omega}\right)\right] \Big/ {\mathcal{A}_j^i} \right\} \tag{40}$$

In the developed game model, different schedulers can receive different payoffs for the same state transition. Based on the TG approach, schedulers seek to choose their power levels self-interestedly to maximize their payoffs. According to the R-learning equation (37), the expected payoff of QoS scheduler:

$$\mathfrak{R}^{\Delta}\left(s,a\right) = U\left(s,a\right) - \eta^{\Delta} + \sum_{s' \in S}\left\{P\left(s' \mid s,a\right) \times \max_{a \in \mathbb{A}} \mathfrak{R}^{\Delta}\left(s',a\right)\right\} \tag{41}$$

In the TG game model, each QoS scheduler is interested in the goal of maximizing his utility function. In a distributed self-regarding fashion, each QoS scheduler in a dynamic IoT system learns the uncertain IoT situation and makes a power control decision by taking into account the online feedback mechanism. With an iterative repeating process, schedulers' decision making mechanism is developed based on the R-learning algorithm, which is an effective way for schedulers' decision mechanism. Based on the dynamic learning mechanism, the developed algorithm can constantly adapt each QoS scheduler's power level to get an appropriate performance balance between contradictory requirements.

Based on the feedback learning process, the RQSC scheme can capture how schedulers adapt their power levels to achieve the better benefit. This procedure is defined as an online power control algorithm. In the RQSC scheme, a selection probability for each power level strategy is dynamically changed based on the payoff ratio, which means the strategy convergence. Therefore, schedulers examine their payoffs periodically in an entirely distributed fashion. Without any impractical rationality assumptions, schedulers can modify their power levels in an effort to maximize their $\mathfrak{N}$ value in a distributed manner. Due to the online self-adjustment technique, this approach can significantly reduce the computational complexity and control overheads. Therefore, it is practical and suitable for the actual system implementation.

In the equation (41), defining of $P\left(s' \mid s,a\right)$ is a probability decision problem. Through the multi-scheduler learning scheme, each scheduler adaptively learns the current IoT system situation to dynamically decide the $P\left(s' \mid s,a\right)$. In the RQSC scheme, each scheduler is assumed to be interconnected by letting them play in a team game with the same environment. Suppose there is a finite set of power levels $\boldsymbol{A}_{i,1\le i\le N}\left(t\right)=\left\{\mathcal{A}_1^i\left(t\right)\ldots\mathcal{A}_m^i\left(t\right)\right\}$ chosen by the scheduler i at the game iteration t. Correspondingly, the $\boldsymbol{U}^i\left(t\right)=\left(u_1^i\left(t\right)\ldots u_m^i\left(\left(t\right)\right)\right.$ is a vector of specifying payoffs for the scheduler i. If the scheduler i plays the action $\mathcal{A}_{l,1\le l\le m}^i$, it earns a payoff $u_{l,1\le l\le m}^i$ with probability p_l^i. Over these actions, $\boldsymbol{P}^i\left(t\right)=\{p_1^i\left(t\right),\ldots,p_m^i\left(t\right)\}$ is defined as the scheduler i's probability distribution.

Power levels chosen by the schedulers are given as input to the environment, and the environmental response to these power levels serves as an input to each scheduler. Therefore, multiple schedulers are connected in a feedback loop with its environment. When a scheduler selects a power level with his respective probability distribution $\boldsymbol{P}\left(\cdot\right)$, the environment produces a payoff $U(\cdot)$ according to the equation (38). Therefore, $\boldsymbol{P}(\cdot)$ should be adjusted adaptively in order to cope with the payoff fluctuation.

At every game round, all schedulers update their probability distributions based on the R-learning algorithm. If the scheduler i chooses $\mathcal{A}_j^i$ at the time t, this scheduler updates the jth action's propensity ($\mathcal{J}_j^i$) as follows;

$$\begin{cases} \mathcal{J}_j^i(t+1) = \mathcal{J}_j^i(t) + \xi_j^i(t), & s.t., \xi_j^i(t) = \left(\left(u_j^i(t) - u_j^i(t-1)\right) \Big/ u_j^i(t-1) \right) \\ \mathcal{J}_k^i(t+1) = \mathcal{J}_k^i(t), & for\ all\ k \neq j \end{cases} \tag{42}$$

The QoS schedulers have to learn an effective action in a distributed fashion while achieving the common objective of IoT system. Known as multi-agent learning approach, the RQSC scheme solves this problem by using distributed R-learning algorithm and docitive paradigm. In the TG model, the main challenge is how to ensure that individual decisions of each QoS scheduler approximate jointly optimal decisions for the team. As a docitive player, individual QoS schedulers cooperate with others by exchanging information while learning the action's propensity from other team members, who are also performing power controls form the R-learning algorithm. In order to apply this approach, QoS schedulers periodically exchange their updated $\mathcal{J}$ values with team members. Based on the received $\mathcal{J}$ values, the jth action's final propensity in the scheduler i ($\mathcal{F}\_\mathcal{J}_{j,1\leq j\leq m}^i(t+1)$) is given by;

$$\mathcal{F}\_\mathcal{J}_j^i(t+1) = \left(\sigma \times \mathcal{J}_j^i(t+1)\right) + \left(\frac{(1-\sigma)}{|\Gamma_j|} \times \sum_{k \in \Gamma_j} \mathcal{J}_j^k(t+1) \right) \tag{43}$$

where σ is a weighted control parameter and Γ_j is the set of schedulers, who send the updated $\mathcal{J}$ values about the jth action for the time $t+1$. Finally, the RQSC scheme can get the $P(s'|s,a)$ value in the equation (41). The ith scheduler's $P(\cdot)$ value for the time $t+1$ ($P^i(S' \mid S, \mathcal{A}_j^i(t+1))$) is defined based on the proportion to each power level's propensity:

$$P^{i}\left(S' \mid S, \mathcal{A}_{j}^{i}\left(t+1\right)\right) = \frac{\mathcal{F}\_\mathcal{J}_{j}^{i}\left(t+1\right)}{\sum_{l=1}^{m} \mathcal{F}\_\mathcal{J}_{l}^{i}\left(t+1\right)} \tag{44}$$

The Main Steps of the RQSC Scheme

In the RQSC scheme, QoS schedulers adaptively decide their power levels while satisfying the QoS needs in coverage areas. Based on past actions and environmental feedback, the RQSC scheme considers the R-learning algorithm and docitive paradigm that attempt to find out optimal actions effectively. Until now, several game models have been developed to help game players to learn from the dynamic network environment. An important feature in the TG model is to enable game players to reach quickly a certain desired game outcome. The main steps of the RQSC scheme can be described as follows:

Step 1: At the initial time, $\mathcal{J}(\cdot)$ is set to be equally distributed ($\mathcal{J}(\cdot)$ = 1/ m where m is the number of power levels), and control parameters σ, N, W, Ω, α, β and ε are given to each QoS scheduler from the simulation scenario (refer to the Table I).

Step 2: At the end of each game's iteration, each QoS scheduler estimates independently its own payoff ($U(\cdot)$) using the equation (38).

Step 3: Based on the currently received information, each QoS scheduler periodically adjusts $\mathcal{J}(\cdot)$ values using the equation (42).

Step 4: According to the docitive paradigm, each QoS scheduler receives the $\mathcal{J}(\cdot)$ values from other QoS schedulers. By using (43), the final propensity for each action is dynamically estimated.

Step 5: Using the proportion to each strategy's propensity, each $P\left(S' \mid S, \mathcal{A}(\cdot)\right)$ is defined according to the equation (44).

Step 6: Iteratively, each QoS scheduler selects a strategy ($\mathcal{A}$) to maximize his long-term expected payoff ($\mathfrak{N}(\cdot)$).

Step 7: The sequential R-learning process is repeatedly operated in a distributed manner.

$$\begin{cases} \mathfrak{N}_{t+1}\left(s,a\right) \leftarrow \mathfrak{N}_{t}\left(s,a\right) \times \left(1-\beta\right) + \beta \times \left(r_{\Theta}\left(s,s'\right) - \eta_{t} + \max_{a' \in A} \mathfrak{N}_{t}\left(s',a'\right)\right) \\ = \mathfrak{N}_{t}\left(s,a\right) + \beta \times \left(r_{\Theta}\left(s,s'\right) - \eta_{t} + \max_{a' \in A} \mathfrak{N}_{t}\left(s',a'\right) - \mathfrak{N}_{t}\left(s,a\right)\right) \\ \eta_{t+1} \leftarrow \eta_{t} \times \left(1-\alpha\right) + \alpha \times \left[r_{\Theta}\left(s,a\right) + \max_{a' \in A} \mathfrak{N}_{t}\left(s',a\right) - \max_{a \in A} \mathfrak{N}_{t}\left(s,a\right)\right] \end{cases} \tag{45}$$

where $r_{\ldots}\left(s,s'\right)$ is the $U\left(\cdot\right)$ from the state s to the state s'

Step 8: If all QoS schedulers reaches the *SE* status, the game process is temporarily stopped. The *SE* status is formally defined as follows.

$$SE = \left\{\mathbb{P} = \left\{P^{i}\left(S' \mid S, \mathcal{A}_{j}^{i}\left(t\right)\right)\right\}_{i \in N, \mathcal{A}_{j}^{i} \in \boldsymbol{A}_{i}} \mid \mathcal{M}\right\}$$

s.t.,

$$\mathcal{M} = \max_{i,\mathcal{A}_{j}^{i}}\left(P^{i}\left(S' \mid S, \mathcal{A}_{j}^{i}\left(t\right)\right) - P^{i}\left(S' \mid S, \mathcal{A}_{j}^{i}\left(t-1\right)\right)\right) \leq \varepsilon \tag{46}$$

Step 9: Constantly, each QoS scheduler is self-monitoring the current IoT situation. If the current system status is not the *SE*, proceeds to Step 2 for the next iteration.

Summary

The RQSC scheme provides a novel QoS control algorithm for IoT systems. Based on the R-learning algorithm and docitive paradigm, the RQSC scheme develops a new TG model. In this game model, QoS schedulers iteratively observe the current IoT system conditions and adaptively change their power levels to maximize the system performance. Due to the self-regarding feature, these control decisions are made by an entirely distributed fashion. In actual IoT system operations, the distributed learning approach is suitable for ultimate practical implementation.

REFERENCES

Akkarajitsakul, K., Hossain, E., & Niyato, D. (2011). Distributed resource allocation in wireless networks under uncertainty and application of Bayesian game. *IEEE Communications Magazine*, *49*(8), 120–127. doi:10.1109/MCOM.2011.5978425

Avenhaus, R. (2004). Applications of inspection games. *Mathematical Modelling and Analysis*, *9*(3), 179–192.

Blasco, P., Giupponi, L., Galindo-Serrano, A., & Dohler, M. (2010). Energy Benefits of Cooperative Docitive over Cognitive Networks.*IEEE Wireless Technology Conference*, (pp. 109-112).

Camerer, C. F. (1997). Progress in Behavioral Game Theory. *The Journal of Economic Perspectives*, *11*(4), 167–188. doi:10.1257/jep.11.4.167

Camerer, C. F. (2003). *Behavioral Game Theory: Experiments in Strategic Interaction*. Princeton University Press.

Camerer, C. F., & Ho, T. (2015). Behavioral Game Theory Experiments and Modeling. Handbook of Game Theory, 4, 517-574.

Camerer, C. F., Ho, T., & Chong, J. (2004a). A Cognitive Hierarchy Model of Games. *The Quarterly Journal of Economics*, *119*(3), 861–898. doi:10.1162/0033553041502225

Camerer, C. F., Ho, T., & Chong, J. (2004b). *Behavioral Game Theory: Thinking, Learning and Teaching*. Paper Presented at the Nobel Symposium 2001, Essays in Honor of Werner Guth.

Chen, K. (2012). Machine-to-Machine Communications for Healthcare. *Journal for Corrosion Science and Engineering*, *6*(2), 119–126.

Dhillon, A., & Mertens, J. (1999). Relative utilitarianism. *Econometrica*, *67*(3), 471–498. doi:10.1111/1468-0262.00033

Dirani, M., & Chahed, T. (2006). Framework for Resource Allocation in Heterogeneous Wireless Networks Using Game Theory. *EuroNGI Workshop*, (pp. 144-154).

Dresher, M. (1962). *A sampling inspection problem in arms control agreements: a game theoretic analysis*. Memorandum RM-2972-ARPA. The RAND Corporation.

Duan, R., Chen, X., & Xing, T. (2011). A QoS Architecture for IOT. *International Conference on and 4th International Conference on Cyber. Physical and Social Computing Internet of Things'2011*, (pp. 717-720). doi:10.1109/iThings/CPSCom.2011.125

Fortino, G., Guerrieri, A., Russo, W., & Savaglio, C. (2014). Integration of agent-based and Cloud Computing for the smart objects-oriented IoT. *IEEE CSCWD, 2014*, 493–498.

Galindo-Serrano, A., & Giupponi, L. (2010). Distributed Q-Learning for Aggregated Interference Control in Cognitive Radio Networks. *IEEE Transactions on Vehicular Technology*, *59*(4), 1823–1834. doi:10.1109/TVT.2010.2043124

Gianini, G., Damiani, E., Mayer, T. R., Coquil, D., Kosch, H., & Brunie, L. (2013). Many-player Inspection Games in Networked Environments. *IEEE DEST*, *2013*, 1–6.

Giluka, M., Kumar, N., Rajoria, N., & Tamma, B. (2014). Class based priority scheduling to support Machine to Machine communications in LTE systems. *IEEE NCC*, (pp. 1-6).

Giupponi, L., Galindo-Serrano, A., Blasco, P., & Dohler, M. (2010). Docitive networks: An emerging paradigm for dynamic spectrum management. *IEEE Wireless Communications*, *17*(4), 47–54. doi:10.1109/MWC.2010.5547921

Ha, V., & Le, B. (2014). Distributed Base Station Association and Power Control for Heterogeneous Cellular Networks. *IEEE Transactions on Vehicular Technology*, *63*(1), 282–296. doi:10.1109/TVT.2013.2273503

He, Q., Rao, Y., & Xu, J. (2010). A Modeling of Game Learning Theory Based on Fairness. *International Journal of Business and Management*, 178–183.

Htike, Z., Hong, C., & Lee, S. (2013). The Life Cycle of the Rendezvous Problem of Cognitive Radio Ad Hoc Networks: A Survey. *Journal for Corrosion Science and Engineering*, *7*(2), 81–88.

Jang, I., Pyeon, D., Kim, S., & Yoon, H. (2013). A Survey on Communication Protocols for Wireless Sensor Networks. *Journal for Corrosion Science and Engineering*, *7*(4), 231–241.

Jiang, C., Chen, Y., Yang, Y., Wang, C., & Liu, K. (2014). Dynamic Chinese Restaurant Game: Theory and Application to Cognitive Radio Networks. *IEEE Transactions on Wireless Communications*, *13*(4), 1960–1973. doi:10.1109/TWC.2014.030314.130632

Kamas, L., & Preston, A. E. (2012). Distributive and reciprocal fairness: What can we learn from the heterogeneity of social preferences? *Journal of Economic Psychology*, *33*(3), 538–553. doi:10.1016/j.joep.2011.12.003

Kim, S. (2011). An Adaptive Online Power Control Scheme based on the Evolutionary Game Theory. *IET Communications*, *5*(18), 2648–2655. doi:10.1049/iet-com.2011.0093

Kim, S. (2012). Reversed Stackelberg bandwidth-sharing game for cognitive multi-hop cellular networks. *IET Communications*, *6*(17), 2907–2913. doi:10.1049/iet-com.2011.0782

Kim, S. (2013). A repeated Bayesian auction game for cognitive radio spectrum sharing scheme. *Computer Communications*, *36*(8), 939–946. doi:10.1016/j.comcom.2013.02.003

Kim, S. (2014). *Game Theory Applications in Network Design*. IGI Global. doi:10.4018/978-1-4666-6050-2

Kinoshita, K., Suzuki, K., & Shimokawa, T. (2013). Evolutionary Foundation of Bounded Rationality in a Financial Market. *IEEE Transactions on Evolutionary Computation*, *17*(4), 528–544. doi:10.1109/TEVC.2012.2208465

Lee, D. (2015). Adaptive Random Access for Cooperative Spectrum Sensing in Cognitive Radio Networks. *IEEE Transactions on Wireless Communications*, *14*(2), 831–840. doi:10.1109/TWC.2014.2360857

Li, L., Li, S., & Zhao, S. (2014). QoS-Aware Scheduling of Services-Oriented Internet of Things. *IEEE Transactions on Industrial Informatics*, *10*(2), 1497–1505. doi:10.1109/TII.2014.2306782

Lien, S., & Che, K. (2011). Massive Access Management for QoS Guarantees in 3GPP Machine-to-Machine Communications. *IEEE Communications Letters*, *15*(3), 311–313. doi:10.1109/LCOMM.2011.011811.101798

Lin, P., Cheng, R., & Liao, L. (2012). Performance Analysis of Two-Level QoS Scheduler for Wireless Backhaul Networks. *IEEE Transactions on Vehicular Technology*, *61*(3), 1361–1371. doi:10.1109/TVT.2012.2186598

Long, C., Zhang, Q., Li, B., Yang, H., & Guan, X. (2007). Non-Cooperative Power Control for Wireless Ad Hoc Networks with Repeated Games. *IEEE Journal on Selected Areas in Communications*, *25*(6), 1101–1112. doi:10.1109/JSAC.2007.070805

Ma, Y., Lv, T., & Lu, Y. (2013). Efficient power control in heterogeneous Femto-Macro cell networks.*IEEE Wireless Communications and Networking Conference (WCNC' 2013)*.

MacKenzie, A. B., & Wicker, S. B. (2001). Game theory in communications: Motivation, explanation, and application to power control. *IEEE GLOBECOM, 2001*, 821–826.

Mahadevan, S. (1996). Average reward reinforcement learning: Foundations, algorithms, and empirical results. *Machine Learning*, *22*(1-3), 159–195. doi:10.1007/BF00114727

Mahapatra, A., Anand, K., & Agrawal, D. P. (2006). QoS and Energy Aware Routing for Real Time Traffic in Wireless Sensor Networks. *Computer Communications*, *29*(4), 437–445. doi:10.1016/j.comcom.2004.12.028

Martin, R., & Tilak, O. (2012). On ε-optimality of the pursuit learning algorithm. *Journal of Applied Probability*, *49*(3), 795–805. doi:10.1017/S0021900200009542

Maschler, M. (1966). A price leadership method for solving the inspector's non-constant sum game. *Naval Research Logistics Quarterly,* 11-33.

Mukherjee, A. (2013). Diffusion of Cooperative Behavior in Decentralized Cognitive Radio Networks With Selfish Spectrum Sensors. *IEEE Journal of Selected Topics in Signal Processing*, *7*(2), 175–183. doi:10.1109/JSTSP.2013.2246136

Nosenzoa, D., Offermanb, T., Seftonc, M., & Veend, A. (2014). Encouraging compliance: Bonuses versus fines in inspection games. *Journal of Law Economics and Organization*, *30*(3), 623–648. doi:10.1093/jleo/ewt001

Pan, Y., Klir, G. J., & Yuan, B. (1996). Bayesian inference based on fuzzy probabilities.*IEEE International Conference on Fuzzy Systems*, *3,* 1693-1699. doi:10.1109/FUZZY.1996.552625

Park, H., & Schaar, M. (2007). Bargaining Strategies for Networked Multimedia Resource Management. *IEEE Transactions on Signal Processing*, *55*(7), 3496–3511. doi:10.1109/TSP.2007.893755

Raazi, S., & Lee, S. (2010). A Survey on Key Management Strategies for Different Applications of Wireless Sensor Networks. *Journal for Corrosion Science and Engineering*, *4*(1), 23–51.

Rabin, M. (1993). Incorporating Fairness into Game Theory and Economics. *The American Economic Review*, 1180–1183.

Savage, L. (1951). The theory of statistical decision. *Journal of the American Statistical Association*, *46*(253), 55–67. doi:10.1080/01621459.1951.10500768

Schwartz, A. (1993). A reinforcement learning method for maximizing undiscounted rewards.*Proceedings of the Tenth International Conference on Machine Learning*, (pp. 298-305). doi:10.1016/B978-1-55860-307-3.50045-9

Singh, D., Tripathi, G., & Jara, A. J. (2014). A survey of Internet-of-Things: Future vision, architecture, challenges and services. *IEEE World Forum on Internet of Things (WF-IoT' 2014)*, (pp. 287-292).

Smith, D. B., Portmann, M., Tan, W., & Tushar, W. (2014). Multi-Source–Destination Distributed Wireless Networks: Pareto-Efficient Dynamic Power Control Game With Rapid Convergence. *IEEE Transactions on Vehicular Technology*, *63*(6), 2744–2754. doi:10.1109/TVT.2013.2294019

Tavoni, A. (2009). *Incorporating Fairness Motives into the Impulse Balance Equilibrium and Quantal Response Equilibrium Concepts: An Application to 2x2 Games*. Fondazione Eni Enrico Mattei - FEEM Working Paper.

Vermesan, O., & Friess, P. (2011). *Internet of Things - Global Technological and Societal Trends*. River Publishers.

Vrancx, P., Verbeeck, K., & Nowé, A. (2008). Decentralized Learning in Markov Games. *IEEE Transactions on Systems, Man, and Cybernetics. Part B, Cybernetics*, *38*(4), 976–981. doi:10.1109/TSMCB.2008.920998 PMID:18632387

Wal, J. (1997). Discounted Markov games: Successive approximation and stopping times. *International Journal of Game Theory*, *6*(1), 11–22.

Xiang, M. (2013). *Trust-based energy aware geographical routing for smart grid communications networks*. Auckland University of Technology. doi:10.1109/TrustCom.2013.12

Xiong, W. (2014). Games under ambiguous payoffs and optimistic attitudes. *Journal of Applied Mathematics*, *2014*, 1–10.

Xiong, W., Luo, X., & Ma, W. (2012). Games with Ambiguous Payoffs and Played by Ambiguity and Regret Minimising Players. *Lecture Notes in Computer Science*, *7691*, 409–420. doi:10.1007/978-3-642-35101-3_35

Xiong, W., Luo, X., Ma, W., & Zhang, M. (2014). Ambiguous games played by players with ambiguity aversion and minimax regret. *Knowledge-Based Systems*, *70*, 167–176. doi:10.1016/j.knosys.2014.06.019

Yu, R., Zhang, Y., Chen, Y., Huang, C., Xiao, Y., & Guizani, M. (2011). Distributed rate and admission control in home M2M networks: A non-cooperative game approach. *IEEE INFOCOM*, (pp. 196-200).

Zhang, Q., & Peng, D. (2012). Intelligent Decision-Making Service Framework Based on QoS Model in the Internet of Things. *International Symposium on Distributed Computing and Applications to Business Engineering and Science*, *2012*, 103–107.

KEY TERMS AND DEFINITIONS

Bayesian Inference: A method of statistical inference in which Bayes' theorem is used to update the probability for a hypothesis as more evidence or information becomes available. Bayesian inference is an important technique in statistics, and especially in mathematical statistics.

Behavioral Game Theory: Analyzes interactive strategic decisions and behavior using the methods of game theory, experimental economics, and experimental psychology. Experiments include testing deviations from typical simplifications of economic theory such as the independence axiom and neglect of altruism, fairness, and framing effects.

Cognitive Hierarchy Theory: A behavioral model originating in behavioral economics and game theory that attempts to describe human thought processes in strategic games. It aims to improve upon the accuracy of predictions made by standard analytic methods, which can deviate considerably from actual experimental outcomes.

Cognitive Radio: An intelligent radio that can be programmed and configured dynamically. Its transceiver is designed to use the best wireless channels in its vicinity. Such a radio automatically detects available channels in wireless spectrum, then accordingly changes its transmission or reception parameters to allow more concurrent wireless communications in a given spectrum band at one location.

CSMA/CA: Carrier sense multiple access with collision avoidance (CSMA/CA) in computer networking, is a network multiple access method in which carrier sensing is used, but nodes attempt to avoid collisions by transmitting only when the channel is sensed to be 'idle'. When they do transmit, nodes transmit their packet data in its entirety.

Power Control: The intelligent selection of transmitter power output in a communication system to achieve good performance within the system.

RFID: Radio-frequency identification (RFID) uses electromagnetic fields to automatically identify and track tags attached to objects. The tags contain electronically stored information. Unlike a barcode, the tag need not be within the line of sight of the reader, so it may be embedded in the tracked object.

Uncertainty: A situation which involves imperfect and/or unknown information. It arises in subtly different ways in a number of fields, including insurance, philosophy, physics, statistics, economics, finance, psychology, sociology, engineering, metrology, and information science.

Chapter 5
Energy-Aware Network Control Approaches

ABSTRACT

Energy is considered as valuable resource for loT network, because the devices used for loT applications are low power-battery operated nodes. In some applications the devices are placed in remote area, when battery of the device would drain out its power, it is difficult to replace the battery. Radio Frequency (RF) energy transfer and harvesting techniques have recently become alternative methods to overcome the barriers that prevent the real world wireless device deployment. Meanwhile, for cellular networks, the base stations (BSs) account for more than 50 percent of the energy consumption of the networks. Therefore, reducing the power consumption of BSs is crucial to energy efficient wireless networks. It can also subsequently reduce the carbon footprints. In this chapter, we focus our attention on the energy-aware IoT control algorithms. For the next-generation IoT systems, they will be key techniques.

DOI: 10.4018/978-1-5225-1952-2.ch005

THE RADIO FREQUENCY BASED ENERGY TRANSFER (RFET) SCHEME

The *Radio Frequency based Energy Transfer* (RFET) scheme is developed as a novel energy harvesting scheme for the Cognitive Radio (CR) network system. Using the sequential game model, data transmission and energy harvesting in each device are dynamically scheduled. The RFET scheme can capture the wireless channel state while considering multiple device interactions. In a distributed manner, individual devices adaptively adjust their decisions based on the current system information while maximizing their payoffs. When data collision occurs, the sequential bargaining process in the RFET scheme can coordinates this problem to optimize social fairness.

Development Motivation

Wireless communication network is becoming more and more important, and has recently attracted a lot of research interest. Compared with wireline communication, wireless communication has lower cost, which is easier to be deployed. With the development of IoT and embedded technology, wireless communication will be applied in more comprehensive scopes. Sometimes, wireless communications work in the license-free band. As a result, it may suffer from heavy interference caused by other networks sharing the same spectrum. In addition, wireless devices perform complex task with portable batteries. However, batteries present several disadvantages like the need to replace and recharge periodically. As the number of electronic devices continues to increase, the continual reliance on batteries can be both cumbersome and costly (Lim, 2013; Park, 2012).

Recently, radio frequency (RF) energy harvesting has been a fast growing topic. The RF energy harvesting is developed as the wireless energy transmission technique for harvesting and recycling the ambient RF energy that is widely broadcasted by many wireless systems such as mobile communication systems, Wi-Fi base stations, wireless routers, wireless sensor networks and wireless portable devices (Khansalee, 2014). Therefore, this technique becomes a promising solution to power energy-constrained wireless networks while allowing the wireless devices to harvest energy from RF signals. In RF energy harvesting, radio signals with frequency range from 300 GHz to as low as 3 kHz are used as a medium to carry energy in a form of electromagnetic radiation. With the increasingly demand of RF energy harvesting/charging, commercialized products, such as Powercaster and Cota system, have been introduced in the market (Lu, 2015).

In wireless communication, Cognitive Radio (CR) technology has evolved to strike a balance between the underutilized Primary User (PU)'s spectrum band and the scarcity of spectrum band due to increased wireless applications. To improve the spectrum utilization efficiency, CR networks make Secondary User (SU) exploit the underutilized PU's band. Nowadays, the CR technology has employed the RF energy harvesting capability that enables SUs to opportunistically not only transmit data on an idle channel, but also harvest RF energy from PUs' transmission on a busy channel (Bhowmick, 2015; Hoang, 2014; Lu, 2014).

Powering a Cognitive Radio Network with RF energy (CRN-RF) system can provide a spectrum-energy efficient solution for wireless communications. In the CRN-RF system, the dual use of RF signals for delivering energy as well as for transporting information has been advocated. Therefore, wireless devices must not only identify spectrum holes for opportunistic data transmission, but also search for occupied spectrum band to harvest RF energy. Such RF signals could be from nearby base stations or access points; they are PUs with energy sources. The RF signal can be converted into DC electricity, and it is stored in an energy storage for the information processing and data transmission. This approach can offer a low-cost option for sustainable operations of wireless systems without hardware modification on the transmitter side. However, due to the specific nature of CRN-RF, traditional CRN protocols may not be directly applied. In addition, the amount of transmitted information and transferred energy cannot be generally maximized at the same time. Therefore, the main challenge in CRN-RF system is to strike a well-balanced trade-off between data transmission and RF energy harvesting (Lu, 2014; Lu, 2015).

Usually, all network agents in the CRN-RF system are assumed to work together in a coordinated manner. However, in practice, network devices are always selfish individuals that will not contribute their work without getting paid. This situation can be seen a game theory paradigm. Under the real-world CRN-RF environments, the system agents are mutually dependent on each other to maximize their payoffs. In the RFET scheme, system agents dynamically adjust their control decisions while responding individually to the current system situations in order to maximize their payoffs. This interactive procedure imitating the sequential game process is practical and suitable for real world CRN-RF implementation. In addition, the RFET scheme considers the social fairness issue among network devices. Therefore, the channel spectrum is shared adaptively according to the sequential bargaining mechanism. In realistic point of view, this approach can be implemented with reasonable complexity.

Energy Harvesting Game Model

To model an energy harvesting game for the CRN-RF system, it is assumed that a cognitive radio network with multiple PUs and SUs; PUs are licensed network agents with energy source, and SUs are unlicensed network devices. For each PU, a non-overlapping spectrum channel is allocated individually, and a channel can be free or occupied by the PU for data transmission. SUs have the RF energy harvesting capability, and perform the channel access by selecting one of them. If the selected channel is busy, the SU can harvest RF energy; the harvested energy is stored in the SU's energy storage. Otherwise, the SU can transmit his data packets (Hoang, 2014).

Simply, the RFET scheme considers a CRN-RF system composed one Cognitive Base Station (CBS), m PUs, and n SUs ($m < n$). Based on the TDMA mechanism, the licensed spectrum channel consists of sequential frames. In other word, each spectrum channel is divided as multiple frames (Niyato, 2014). Frames over time are used for data transmission or RF energy transferring. Under dynamic network environments, the PU can use the whole frame for his own data transmission, or vacates his frames. When a frame is actively used for the PU's data transmission, the RF energy can be transferred to the SUs through the RF energy harvesting process. If the frame is vacated, this vacated frame can be used by a SU. The illustrative structure of channel frames is shown in the Figure 1.

The RFET scheme develops a new energy harvesting game model ($\mathbb{G}$) for the CRN-RF system. To model strategic CRN-RF situations, the RFET scheme assumes that unlicensed network devices, i.e., SUs, are game players. During the interactive sequential game process, players choose their strategy based on the reciprocal relationship. From the view of individual SUs, the main challenge is to effectively transmit his data or to harvest RF energy. Therefore, the game $\mathbb{G}$ is designed as a symmetric game with the same strategy set for game players.

Figure 1. Spectrum channel structure with multiple frames

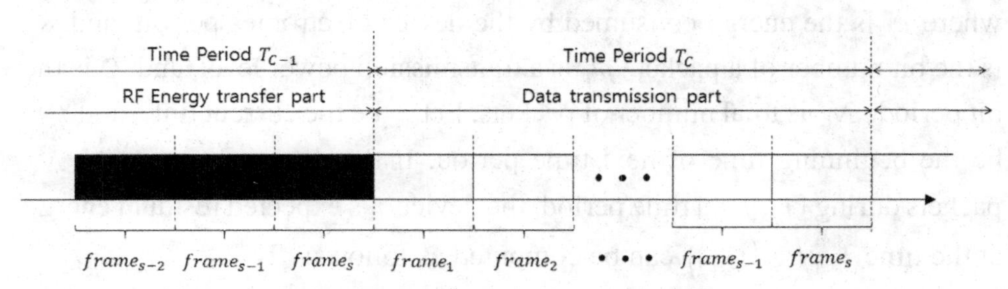

Definition 1: The energy harvesting game model constitutes three game tuple $\mathbb{G} = (\boldsymbol{N}, \boldsymbol{S}_{i,0 \le i \le n}, \boldsymbol{U}_{i,0 \le i \le n})$, where

1. $\boldsymbol{N}$ is a set of game players; $i \in \boldsymbol{N} = \{1,\ldots,n\}$ is an unlicensed network device in the CRN-RF system,
2. $\boldsymbol{S}_i = \{s_i^1,\ldots s_i^k \ldots s_i^m\}$ is a nonempty finite strategy set of the player $i \in \boldsymbol{N}$; s_i^k means that the kth PU channel is selected by the player i for the CRN-RF service.
3. $\boldsymbol{U}_i$ is the utility function to represent the payoff of player $i \in \boldsymbol{N}$. $\boldsymbol{U}_i$ is decided according to the set of all players' strategies: ($s_1^{(\cdot)} \times \ldots s_i^{(\cdot)} \ldots \times s_n^{(\cdot)}) \rightarrow \boldsymbol{U}_i$.

In the game model $\mathbb{G}$, each PUs arbitrarily use their channels to transmit their data over time. Therefore, SUs can temporally access the PUs' channels in an opportunistic manner. As the data sender or RF energy harvester, the major goal of SUs is to jointly optimize the data transmission and energy harvesting. If SUs wants to transmit their data, they try to find the vacant PU channels. If multiple SUs access a specific vacant channel at the same time, the channel frames are adaptively distributed for each SU to maximize the total CRN-RF system performance. If the PU come back to use his designated channel, SUs should release the momentary-using channel and try to find other idle channels (Hoang, 2014; Lu, 2015). If SUs wants to harvest the energy, they try to find the active data transferring channels of PUs.

The RFET scheme assumes a practical energy consumption and harvesting model for wireless networks. For a packet transmitted from a transmitter device to a receiver device, the energy (E_t) consumed at the packet transmitter is defined as (Feng, 2009);

$$E_t\left(\mathfrak{F}_n, \mathcal{N}_p\right) = \left(\left(e_e \times \mathfrak{F}_n\right) + \left(pl_t \times \mathcal{L}\right)\right) \times \mathcal{N}_p \tag{1}$$

where e_e is the energy consumed by the device electronics per bit, and $\mathfrak{F}_n$ is the bit number of a packet. pl_t is a transmission power level, and $\mathcal{L}$ is the bit period. $\mathcal{N}_p$ is total number of packets. Let t_c be the current time, and t_{c+1} be the beginning time of next time period. If the device i transmits $\mathcal{N}_p^i$ packets during [t_c, t_{c+1}] time period, the device i's expected residual energy at the time t_{c+1} $\left(\Phi_i\left(t_{c+1}\right)\right)$ can be computed as follows:

$$\Phi_i\left(t_{c+1}\right) = \Re_i\left(t_c\right) - E_t\left(\mathfrak{F}_n, \mathcal{N}_p^i\right) \tag{2}$$

where $\Re_i\left(t_c\right)$ is the device i's residual energy level at time t_c. In contrast to the energy consumption, the energy harvesting is the process by which energy is derived from external sources, and stored in wireless devices. If a PU actively transmits his own data during the time period [t_c, t_{c+1}], the RF energy harvesting rate ($\varepsilon^H\left(t_c, t_{c+1}, d\right)$) in SUs is given as follows (Aslam, 2015; Feng, 2009);

$$\varepsilon^H\left(t_c, t_{c+1}, d\right) = \int_{t=t_c}^{t=t_{c+1}} \left(\Gamma \times p_t \times \left(\frac{G_t \times G_r}{\left(4\pi\right)^2 \times d^\alpha}\right)\right) dt \tag{3}$$

where Γ is a factor of energy harvesting efficiency, and p_t is the signal transmission power of the PU with RF resource. G_t and G_r are the gains of the energy transmitter and receiver, respectively. d is the distance between the RF source and the energy receiving device, and α is the path-loss exponent.

Fundamental problem in game theory is determining how players reach their decisions. Usually, players selfishly select a strategy that maximizes their own payoff through the utility function. To design the utility function (U_i) of player i, the RFET scheme should define the received benefit and the incurred cost during the CRN-RF services. Let $\beta_{i,k}\left(pl_t^i\right)$ be the data transmission amount that the player i can achieve over the kth frame in a channel with the power level pl_t^i. Using Shannon formula, the player i's time-varying data transmission amount can be computed as follows:

$$\beta_{i,k}\left(pl_t^i\right) = W_k \times \left(\log_2\left(1 + \gamma_{i,k}\left(pl_t^i\right)\right)\right) \tag{4}$$

where, W_k is the bandwidth of kth frame and $\gamma_{i,k}(pl_t^i)$ is the signal to interference and noise ratio (SINR) over the kth frame (Aslam, 2015). The cost is defined as the instantaneous expense function of consumed energy. Finally, the player i's utility function ($U_i\left(\cdot\right)$) during [t_c, t_{c+1}] is defined as follows;

$$U_i\left(pl_t^i, t_c, t_{c+1}\right) = \begin{cases} \sum_{k \in \mathbb{S}_i^{t_c, t_{c+1}}} \left\{ \left(\beta_{i,k}\left(pl_t^i\right)\right) - \xi \times \left(E_t^i\left(\mathfrak{F}_n, \mathcal{N}_p^{i,k}\right)\right)^q \right\}, & \textit{if data transmission strategy is selected} \\ \varepsilon^H\left(t_c, t_{c+1}, d\right), & \textit{if energy harvesting strategy is selected} \end{cases} \quad (5)$$

where $\mathbb{S}_i^{t_c, t_{c+1}}$ is the set of frames, which are allocated for the player i during $[t_c, t_{c+1}]$. pl_t^i and $E_t^i(\cdot)$ are the power level and the E_t value of the player i, respectively. ξ and q are cost parameters. $\mathcal{N}_p^{i,k}$ are the total number of packets in the kth frame of the player i.

Cognitive Radio Sensing Mechanism

Usually, allocated spectrum channels are largely unused in any time and location; these are referred to as spectrum holes. The increased spectrum utilization in CR networks is achieved through opportunistically sharing the spectrum holes between licensed and unlicensed wireless devices (Kim, 2014). To detect spectrum holes, unlicensed wireless devices are sensing constantly the allocated spectrum channel. Based on the received signal from the detection, the RFET scheme can define two hypotheses for the case of the licensed device is present (H_1) or absent (H_0). If there are K detection samplings, the kth received signal for the channel l (r_l^k, i.e., $1 \leq k \leq K$) can be written as

$$r_l^k = \begin{cases} \left(h^k \times s\left(k,l\right)\right) + w\left(k,l\right), & if\ H_1 \\ w\left(k,l\right), & if\ H_0 \end{cases} \quad (6)$$

where h^k is the channel gain from the licensed wireless device to the unlicensed wireless device; it is assumed to be slow flat fading. $s(k,l)$ is the channel l's signal of the licensed wireless device, and $w(k,l)$ is the channel l's additive white Gaussian noise (AWGN) with mean zero and variance σ_w^2 (Wang, 2010). In the RFET scheme, $s\left(k,l\right)$ and $w(k,l)$ are assumed to be mutually independent. In CR systems, each unlicensed wireless device senses the channel and sends the sensing information r to the CBS. Based on the r_l^k information, the CBS defines the channel l's test statistics $\left(\mathfrak{N}\left(l\right)\right)$ as follows;

$$\mathfrak{N}\left(l\right)=\frac{1}{K}\times\left(\sum_{k=1}^{K}\left(r_l^k\right)^2\right) \tag{7}$$

where K is the total number of collected samples (Wang, 2010). The CBS maintains a look-up vector ($\mathbb{V}$) for each channels. In the RFET scheme, $\mathbb{V}$ represent the $\mathfrak{N}\left(\cdot\right)$ outcomes of the last x time rounds [t_{c-x}, t_{c-1}] while emphasizing the most recent information. By using an appropriate time-oriented fusion rule, the values in $\mathbb{V}$ is estimated in the timed weighted manner. At time t_c, the $\mathbb{V}$ value for the channel l ($\mathbb{V}^{t_c}\left[l\right]$) is given by;

$$\mathbb{V}^{t_c}\left[l\right]=\sum_{f=t_{c-x}}^{t_{c-1}}\left(\left(1-\mu\right)^{t_c-f}\times\mathfrak{N}^f\left[l\right]\right) \tag{8}$$

where μ is a control factor, which can control the weight for each timed strategies, and $\mathfrak{N}^f\left[l\right]$ is the $\mathfrak{N}\left[l\right]$ at the time f. Based on the $\mathbb{V}$ [•] information, game players individually decide which PU's channel is selected to maximize their payoffs. To adaptively make these decisions, each player has a data queue ($\mathbb{Q}$) and energy storage ($\mathbb{E}$). $\mathbb{Q}$ is used to store the generated data for transmission, and $\mathbb{E}$ is a battery to store RF energy harvested from radio signal (Hoang, 2014). $\mathfrak{T}$ and $\mathfrak{P}$ ($0\leq\mathfrak{T}\leq\mathcal{M}^{\mathbb{Q}}$ and $0\leq\mathfrak{P}\leq\mathcal{M}^{\mathbb{E}}$) represent the packet amount in the $\mathbb{Q}$, and energy level of the $\mathbb{E}$, respectively; $\mathcal{M}^{\mathbb{Q}}$ is the maximum size of $\mathbb{Q}$ and $\mathcal{M}^{\mathbb{E}}$ is the maximum capacity of $\mathbb{E}$.

Based on the $\mathbb{V}$ [•], $\mathfrak{T}$ and $\mathfrak{P}$ values, each player selects his strategy whether to transmit data or to harvest energy. To formulate the player i's channel selection problem at the time t_c, let $\eta_j^i(t_c)$ is the player i's propensity for the j's channel selection. Based on the sophisticated combination of the remaining energy and data arriving rate, $\eta_j^i(t_c)$ is dynamically estimated for each individual player.

$$\eta_j^i(t_c) = \left[\frac{\mathfrak{T}}{\mathcal{M}^{\mathbb{Q}}} \times \left((\mathfrak{M} - \mathbb{V}[j]) \times \left(\sum_{l=1}^{M} (\mathfrak{M} - \mathbb{V}[l]) \right)^{-1} \right) \right] + \left[\left(\frac{\mathcal{M}^{\mathbb{E}} - \mathfrak{P}}{\mathcal{M}^{\mathbb{E}}} \right) \times \left(\mathbb{V}[j] \times \left(\sum_{l=1}^{m} \mathbb{V}[l] \right)^{-1} \right) \right],$$
$$s.t., \mathfrak{M} = \max_{1 \le l \le M} \mathbb{V}[l] \quad (9)$$

where m is total number of PU channels. According to (9), the channel j's selection probability ($\mathcal{P}_j^i$) from the player i at time t_c can be captured as follows;

$$\mathcal{P}_j^i(t_c) = \eta_j^i(t_c) / \left(\sum_{l=1}^{m} \eta_l^i(t_c) \right) \quad (10)$$

For the data transmission, individual players try to select a vacant PU channel. If multiple players select the same vacant PU channel, the RFET scheme should decide how to share this licensed spectrum channel to multiple SU. To solve this problem, the RFET scheme adopts the Sequential Bargaining Solution (SBS), which can combine the bargaining problem with social fairness. During continuative time period, SBS process can occur sequentially to capture the strategic interaction among players (Wu, 2014). Based on the player's $\mathfrak{T}$ value, each player tradeoffs to yield the maximum social benefit. Let $\mathcal{Q} = \{1 \dots i \dots r\} \subseteq N$ be the set of players selecting the same channel, the SBS at time t_c $\left(\boldsymbol{SBS}(t_c)\right)$ is $\partial(t_c) = \{\partial_1^{t_c} \dots \partial_i^{t_c} \dots \partial_r^{t_c}\}$ where $\partial_i^{t_c}$ represents the amount of frames assigned to the player i during [t_c, t_{c+1}]. The $\boldsymbol{SBS}(t_c)$ is obtained through the following maximization problem;

$$\boldsymbol{SBS}(t_c) = \max_{\partial(t_c) = \{\partial_1^{t_c} \dots \partial_i^{t_c} \dots \partial_r^{t_c}\}} \sum_{i \in \mathcal{Q}} \left(Y_i \times \log_2 \left(1 + U_i \left(pl_t^i, t_c, t_{c+1} \right) \right) \right)$$

s.t.,

$$\sum_{i \in \mathcal{Q}} Y_i = 1 \text{ and } Y_i = \mathfrak{T}_i / \sum_{i \in \mathcal{Q}} \mathfrak{T}_i \quad (11)$$

where Y_j is the bargaining power of the player j; it is the relative ability to exert influence for the bargaining solution. In the RFET scheme, the bargaining powers are decided according to the players' $\mathfrak{T}$ values. If a player has a relatively higher data amount in his $\mathbb{Q}$, he has a higher bargaining power. Therefore, over the time period, the players' $\mathfrak{T}$ values can be balanced while ensuring social fairness among players.

Summary

Spectrum efficiency and energy efficiency are two critical issues in designing wireless networks. As energy harvesting becomes technologically viable, RF energy harvesting has emerged as a promising technique to supply energy to wireless network devices. On the other hand, we can improve the spectrum efficiency and capacity through CR spectrum access. Currently, it has raised a demand for developing new control protocols to maximize the wireless information transferring and energy harvesting simultaneously. The RFET scheme designs a new energy harvesting algorithm for RF-CRN systems. Focusing on the trade-off between data transmission and RF energy harvesting, spectrum channels are dynamically selected and adaptively shared by unlicensed network devices. Based on the iterative game model and sequential bargaining process, the RFET scheme can achieve an effective RF-CRN system performance.

THE FAIR EFFICIENT ENERGY CONTROL (FEEC) SCHEME

In the recent decades, cellular networks have revolutionized the way of next generation communication networks. However, due to global climate changes, reducing the energy consumption of cellular infrastructures is an important and urgent problem. The Fair Efficient Energy Control (FEEC) scheme is a novel two-level cooperative game framework for improving the energy efficiency and fairness in cellular networks. For the energy-efficiency, base stations (BSs) constantly monitor the current traffic load and cooperate each other to maximize the energy saving. For the energy-fairness, renewable energy is shared dynamically while guaranteeing fairness among BSs. In the dual-level cooperative game model in the FEEC scheme, the concept of the *Raiffa bargaining Solution* and *Jain's fairness* are extended and practically applied to achieve an excellent cellular network performance.

Development Motivation

The current explosive popularity of smartphones and mobile devices has ignited a surging traffic demand for wireless accesses and has been incurring massive energy consumption, which causes global warming due to CO_2 emissions. With the increasing awareness of global warming and environmental consequences of Information and Communications Technology (ICT), researchers have been seeking ways to reduce energy consumption. As a significant component of ICT energy consumption, cellular networks will have greater economic and ecological impact in the coming years. Concentrating of environmental influences, a new research area called '*green cellular networks*' has recently emerged to enable various energy-efficient cellular networks. The *green* approach can gain extra commercial benefits, mainly by reducing operating expense related to energy cost (Wu, 2015; Zheng, 2015).

In wireless cellular networks, energy consumption is mainly drawn from base stations (BSs); they account for more than 50 percent of the energy consumption of the cellular networks. In addition, the number of BSs is expected to be doubled by 2012 (Han, 2013). Therefore, promising technology has been developed to improve the energy efficiency of BSs; it is crucial to achieve green cellular networks while approximating an optimal energy saving. From the perspective of cellular network management, energy-efficient operation of BSs is not only a matter of social environmental responsibility, but also tightly related to cellular network control issues (Chia, 2014; Han, 2013; Reyhanian, 2015).

The main focus of FEEC scheme is devoted to maximize the energy efficiency in BSs. To address the challenge of increasing energy efficiency and profitability in future green cellular networks, it is important to consider various paradigm-shifting methodologies (Han, 2013; Zheng, 2015). Among all the promising energy saving approaches, BS sleeping and renewable energy (RE) distribution methods are very effective, and prominent solutions to optimize the energy utilization in green cellular networks (Chia, 2014; Han, 2013; Reyhanian, 2015). The past decade has seen a surge of research activities in each individual method. However, few research has been done for the proper combination of these two methods.

The intuitive idea of BS sleeping method is to switch off the BSs when the traffic load is below a certain threshold for a certain time period. Simply, the BS sleeping problem can be formulated as an optimization problem that minimizes the number of active BSs while meeting the traffic load in cellular networks. This problem is a well-known combinatorial optimization problem (Han, 2013). However, the BS sleeping problem has been proven to be NP-hard. Moreover, solving this kind of problem generally requires a central controller as well as the global information, which makes the problem more challenging (Zheng, 2015).

Another leading approach for green cellular networks is the use of RE (Chia, 2014; Reyhanian, 2015). In recent years, RE resources, such as solar panels and wind turbines, can be considered as complementary energy providers for conventional power systems, such as diesel generators or the power grid. Due to the significantly higher cost of conventional power generation, RE becomes more and more attractive. Although RE sources are attractive for green cellular network operations, they suffer from higher variability as compared to conventional energy sources (Chia, 2014). Therefore, a key consideration for the RE management is to maximize the energy efficiency while compensating the variability of the RE sources.

Under widely dynamic cellular system conditions, BSs and RE Providers (REP) can be assumed as intelligent rational decision-makers, and they select a best-response strategy to maximize their expected payoffs. This situation is well-suited for the game theory. Therefore, game theory is really useful in analyzing the mutual interactions among BSs and REPs. Thus, it can be a major paradigm to retain an effective solution that feature the complex interactive relations of BSs and REPs in green cellular networks.

To ensure fair-efficient energy provisioning, the methodology adopted in the FEEC scheme is a two-level cooperative game model. This approach can improve effectively the energy efficiency and fairness of cellular networks. At the efficiency-control stage, some BSs are grouped together as a cluster, and work together toward an optimal system performance. When the traffic load of a specific BS is below a certain threshold, this BS is switched off, and running services are taken care of by the neighboring BSs that remain active in the cluster. At the fairness-control stage, REPs fairly distribute the available RE to their corresponding BSs. Under dynamically changing network environments, BSs and REPs coordinate with each other in order to ensure the fair-efficient energy management. Therefore, the dual-level game approach is suitable to get a globally desirable network performance.

Two-Level Game Model for Green Cellular Networks

The ever-increasing traffic demand in cellular networks has triggered a large expansion of network infrastructures and BSs, which results in substantial increase in energy consumption. At present, all BSs in cellular networks are working on the 'always-ON' state, regardless of the traffic levels associated with them. Moreover, a traditional on-state BS is designed to satisfy peak traffic requirements. However, in fact, the worldwide average peak utilization rates of BSs are merely at 65% ~ 70%. This situation motivates many power saving efforts that have been done towards under-utilized BSs. In particular, BS sleeping strategy is an effective mechanism to reduce energy consumption of cellular networks. Generally, a sleeping BS reduces its energy consumption by 1/2 ~ 2/3 compared with its active mode. Therefore, it is reasonable to let a subset of BSs go to sleep when the traffic load is below a certain threshold (Bao, 2015; Zhang, 2013).

During the cellular networks system operations, system agents, such as BSs and REPs, should make decisions individually. In this situation, a main issue for each agent is how to perform well by considering the mutual-interactive relationship while dynamically adjusting their decisions to maximize their own profits. In the FEEC scheme, the dynamic operation of BSs and REPs is formulated as a two-level game model. At the first stage, BSs play an efficiency-control level game; the BSs dynamically create an ad-hoc cluster, and a low traffic load BS is switched off. At this moment, the traffic load in the sleeping BS is effectively shared through a cooperative manner. At the second stage, BSs and REP play a fairness-control level game; some BSs are grouped as a permanent cluster with one REP, and each REP provides the available RE to its corresponding BSs through a cooperative manner. For the implementation practicality, the FEEC scheme is designed in an entirely distributed and self-organizing interactive fashion.

Mathematically, the efficiency-control level game ($\mathbb{G}^{EC}$) can be defined as

$$\mathbb{G}^{EC} = \left\{ \mathbb{N}, \mathcal{C}_{i,i\in\mathbb{N}}, \mathbb{L}, \left\{ \boldsymbol{S}_j \right\}_{j\in\mathbb{N}}, \left\{ U_j \right\}_{j\in\mathbb{N}}, T \right\}$$

at each time period t of gameplay.

- $\mathbb{N} = \{1, \ldots, n\}$ is the finite set of all BSs; they are game players.
- $\mathcal{C}_i$ is the ad-hoc cluster for the sleeping BS i, i.e., $i \in \mathbb{N}$. Multiple ad-hoc clusters can exist.
- $\mathbb{L}$ is the set of BSs' traffic loads in the $\mathcal{C}$; $\mathbb{L} = \{l_1, l_2, \ldots, l_m\}$ where m is the number of BSs in the $\mathcal{C}$.
- $\boldsymbol{S}_j$ is the set of strategies with the BS $j \in \mathcal{C}$. $\boldsymbol{S}_j$ represents the amount of taken traffic load from the sleeping BS in $\mathcal{C}$.

- U_j is the payoff received by the BS j. It is the profit obtained from the BS sleeping algorithm.
- The T is a time period. The $\mathbb{G}^{EC}$ is repeated $t \in T < \infty$ time periods with imperfect information.

To distribute the RE to BSs, BSs in green cellular networks are also clustered permanently. One REP has its own permanent cluster, and each REP is responsible to distribute the RE to the BSs in its corresponding cluster. To formulate interactions between BSs and REPs, the fairness-control game ($\mathbb{G}^{FC}$) can be defined as

$$\mathbb{G}^{FC} = \left\{ \mathbb{K}, \left\{ \mathcal{M}^i \right\}^{i \in \mathbb{K}}, \left\{ \mathbb{A}^i \right\}^{i \in \mathbb{K}}, \left\{ \mathcal{U}_j^i \right\}_{j \in \mathcal{M}^i}^{i \in \mathbb{K}}, T \right\}$$

at each time period t of gameplay.

- $\mathbb{K}$ is the finite set of REPs in the green cellular network; $\mathbb{K} = \left\{ \mathcal{K}_1, \ldots, \mathcal{K}_z \right\}$ where z is the total number of REP, and $\mathcal{K}_{i, 1 \leq i \leq z}$ is responsible to the permanent cluster i.
- $\mathcal{M}^i = \left\{ \mathcal{B}_1^i, \ldots, \mathcal{B}_w^i \right\}$ is the finite set of BSs in the permanent cluster i where w is the number of BSs in the $\mathcal{M}^i$. In the green cellular architecture, $z \times w$ is equal to $n = \left| \mathbb{N} \right|$, and $\left\{ \mathbb{K}, \left\{ \mathcal{M}^i \right\}^{i \in \mathbb{K}} \right\}$ are game players of $\mathbb{G}^{FC}$.
- $\mathbb{A}^i = (\mathcal{A}_1^i, \mathcal{A}_2^i, \ldots, \mathcal{A}_w^i)$ is the set of strategies for the $\mathcal{M}^i$; $\mathbb{A}^i$ represents the amount of allocated RE, and it is decided by the REP $\mathcal{K}_i \in \mathbb{K}$.
- The $\mathcal{U}_j^i$ is the payoff received by the BS $j \in \mathcal{M}^i$. It is the BS j's profit obtained from the REP $\mathcal{K}_i$.
- The T is a time period. The $\mathbb{G}^{FC}$ is repeated $t \in T < \infty$ time periods with imperfect information.

Efficiency-Control Game at the First Stage

In the FEEC scheme, a cellular network deployed in an urban area is considered, and each cell may experience various traffic densities over space and time. Each cell is typically shaped as a hexagonal area, and it is serviced by a BS. The FEEC scheme assumes that each of BSs belongs to a rational operator. For adaptive power control decisions, each BS exchanges local information with the adjacent neighboring BSs periodically. Therefore, BS cooperation has the potential to reduce effectively energy consumption in cellular networks. To save the consuming energy, the FEEC scheme can switch off the BS with light traffic load while transferring the running services to its neighboring BSs. Due to its capability, this approach will become more popular in the future (Bao, 2015).

In general, the power consumption of BS j ($\Psi_j^{\mathfrak{B}}$) is composed of two types of power consumptions: fixed power consumption ($\mathcal{F}_j^{\mathfrak{B}}$) and dynamic power consumption ($\mathcal{D}_j^{\mathfrak{B}}$). $\mathcal{F}_j^{\mathfrak{B}}$ denotes the power consumed statically even though a BS is idle. In active BS mode, it includes power cost by power amplifier, feeder, transmit antennas and air conditioning. $\mathcal{D}_j^{\mathfrak{B}}$ mainly denotes the power used for actual data transmission; it is related to the current traffic load of BS j (Zhang, 2013; Zheng, 2015). For the case of BS sleeping, BS switching cost ($\mathcal{S}^{\mathfrak{B}}$) occurs; it is the extra power consumption when the status of BS transforms between active and sleep modes. Depending on the associated traffic load, $\mathcal{S}^{\mathfrak{B}}$ is estimated by including the traffic transferring overhead.

In the FEEC scheme, each BS announces its own traffic load (l) periodically to the Mobile Switching Center (MSC), which works as a network gateway, and is responsible for the inter-cell management of BSs. When the traffic load of BS i is less than the threshold (Γ), the BS i and its neighboring BSs dynamically form an ad-hoc cluster ($\mathcal{C}_i$). And then, the BS i attempts to transfer its traffic load to the neighboring active BSs in $\mathcal{C}_i$. In the $\mathcal{C}_i$, the transferring traffic load ($\hat{l}$) from the sleeping BS i is denoted by

$$\hat{L}_{-i} = \left[\widehat{l_1^i}, \widehat{l_2^i} \ldots \widehat{l_{i-1}^i}, \widehat{l_{i+1}^i}, \ldots, \widehat{l_m^i}\right];$$

$\hat{L}_{-i}$ is decided geographically. If the BS i is switched off, the active BSs in the $\mathcal{C}_i$ will get paid according to the $\hat{L}_{-i}$. Therefore, the BS sleeping problem can be formulated to an optimization problem as follows; the goal of this problem is to minimize the power consumption of $\mathcal{C}_i$.

$$\min_{S_j}\left(j \in \mathcal{C}_i \subset \mathbb{N} \mid \sum_j \Psi_j^{\mathfrak{B}}\right)$$

s.t.,

$$\begin{cases} if\ BS\ i\ is\ not\ switched\ off, & \Psi_j^{\mathfrak{B}} = \mathcal{F}_j^{\mathfrak{B}} + \mathcal{D}_j^{\mathfrak{B}}, where\ all\ j \in \mathcal{C}_i \\ if\ BS\ i\ is\ switched\ off, & \begin{cases} \Psi_j^{\mathfrak{B}} = \mathcal{F}_j^{\mathfrak{B}} + \mathcal{D}_j^{\mathfrak{B}} + \mathfrak{T}\left(\hat{l}_j\right), where\ j \in \mathcal{C}_i - \{i\} \\ \Psi_i^{\mathfrak{B}} = \mathcal{S}_i^{\mathfrak{B}}, where\ i \neq j \end{cases} \end{cases}$$

and

$$\mathcal{D}_j^{\mathfrak{B}} = \eta \times l_j, \mathcal{S}_i^{\mathfrak{B}} = \varepsilon \times l_i,\ \mathfrak{T}\left(\hat{l}_j\right) = \left(\varepsilon \times \hat{l}_j\right) + \left(\eta \times \hat{l}_j\right) \tag{12}$$

where η and ε are the energy parameters for the traffic load (l) execution and transferring, respectively. Based on the Eq. (12), the FEEC scheme adaptively solves the BS sleeping problem in order to minimize the power consumption in $\mathcal{C}_i$.

Even though the BS sleeping strategy is essential to the energy efficiency of green cellular networks, frequent BS on/off switching may cause the degradation to the Quality-of-Service (QoS) while increasing the network operational cost. To avoid the frequent mode transitions, the FEEC scheme develops a dual-threshold based sleep mechanism. When the traffic load of BS is less than the lower threshold (Γ_L), the BS will switch off. When the traffic load of BS reaches the upper threshold (Γ_H), the BS will switch on. This dual-threshold approach can effectively prevent shuttling of BS status between on and off states. Based on the state transition cost and current power consumption, the FEEC scheme adaptively adjusts two threshold values while minimize the energy consumption. Therefore, the Γ_L and Γ_H of BS i are defined as follows.

$$\begin{cases} \Gamma_L^i = \mathcal{S}_i^{\mathfrak{B}} \\ \Gamma_H^i = \dfrac{1}{\left|\mathcal{C}_i\right|} \times \sum_{j \in \mathcal{C}_i} l_j \end{cases} \tag{13}$$

If the active BS i's $\Psi_i^{\mathfrak{B}}$ is less than Γ_L^i, the BS i is switched of If the traffic load in the sleeping BS i's area reaches the Γ_H^i, the BS i will switch on. In the FEEC scheme, the MSC and neighboring BSs are assumed to have the ability of detecting the traffic load in the sleeping BS's area.

To design the BS sleeping algorithm, the FEEC scheme should consider how self-interested BSs would agree to serve the traffic load in the sleeping BS's area while using their resources. The FEEC scheme uses an incentive-payment ($\mathfrak{P}$) technique because it can make a self-organizing system effectively functional. For neighboring BSs, the incentive-payment

$$\mathfrak{P} = \left[\mathfrak{P}_1(\cdot)\ldots\mathfrak{P}_{i-1}(\cdot),\mathfrak{P}_{i+1}(\cdot)\ldots,\mathfrak{P}_m(\cdot)\right]$$

is provided to induce selfish BSs to participate in the BS sleeping mechanism. To ensure the socially efficient outcome, incentive compatibility, budget balance and participation constraints, $\mathfrak{P}$ should be dynamically decided. Based on the $\mathfrak{P}$, the neighboring BS j's utility function $\left(U_j(\cdot)\right)$ is defined by

$$U_j\left(\widehat{l_j^i},\widehat{\Psi_j^{\mathfrak{B}}}\right) = -\left[\left(\Delta_e \times \left(\eta \times \widehat{l_j^i}\right)\right) + \left(\Delta_t \times \left(\varepsilon \times \widehat{l_j^i}\right)\right)\right] + \mathfrak{P}_j\left(\hat{L}_{-i},\widehat{\Psi_j^{\mathfrak{B}}}\right)$$

s.t.,

$$\hat{L}_{-i} = \left[\widehat{l_1^i},\widehat{l_2^i}\ldots\widehat{l_{i-1}^i},\widehat{l_{i+1}^i},\ldots,\widehat{l_m^i}\right], i \neq j \text{ and } i,j \in \mathcal{C}_i \quad (14)$$

where $\widehat{l_j^i}$ is the transferring traffic load from the sleeping BS i to the BS j, and $\widehat{\Psi_j^{\mathfrak{B}}}$ is the extra power consumption for the $\widehat{l_j^i}$. Δ_e, Δ_t are the cost fee parameters for using electricity, and transferring overhead for the traffic load ($\widehat{l_j^i}$), respectively. $\mathfrak{P}_j\left(\hat{L}_{-i},\widehat{\Psi_j^{\mathfrak{B}}}\right)$ is decided by the MSC based on the traffic distribution information $\hat{L}_{-i}$. According to the rational participation constraint, $\mathfrak{P}_j(\cdot)$ is dynamically decided to guarantee $U_j(\cdot)\geq 0$. This condition can translate the selfish motives of BSs into desirable actions for traffic sharing.

The FEEC scheme develops a cooperative game mechanism to decide $\mathfrak{P}$. The main design goal is to effectively re-distribute the saving energy while meeting the rational constraints. To satisfy this goal, the FEEC scheme adopts the concept of *Raiffa Bargaining Solution* (*RBS*); this solution can ensure the *Pareto Optimality*, *Independence of Linear Transformations*, *Symmetry*, and *Monotonicity* (Bao, 2015). To implement the RBS, the BS j's preference function $v_j\left(\hat{L}_{-i},\widehat{\Psi_j^{\mathfrak{B}}}\right)$ is defined with the minimum utility $U_j^{\min}\left(\hat{l_j^i},\widehat{\Psi_j^{\mathfrak{B}}}\right)$ and maximum utility $U_j^{\max}\left(\hat{l_j^i},\widehat{\Psi_j^{\mathfrak{B}}}\right)$:

$$v_j\left(\hat{L}_{-i},\widehat{\Psi_j^{\mathfrak{B}}}\right)=\begin{pmatrix}\left(U_j\left(\hat{l_j^i},\widehat{\Psi_j^{\mathfrak{B}}}\right)-U_j^{\min}\left(\hat{l_j^i},\widehat{\Psi_j^{\mathfrak{B}}}\right)\right)+\frac{1}{\left|\mathcal{C}_i\right|-2}\\ \times\left(\sum_{k\in\mathcal{C}_i,k\neq i,j}U_k^{\max}\left(\hat{l_k^i},\widehat{\Psi_j^{\mathfrak{B}}}\right)-U_j\left(\hat{l_j^i},\widehat{\Psi_j^{\mathfrak{B}}}\right)\right)\end{pmatrix}^{\alpha_j},s.t.,\sum_{j\in\mathcal{C}_i,j\neq i}\alpha_j=1 \quad (15)$$

where $U_j^{\min}\left(\hat{l_j^i},\widehat{\Psi_j^{\mathfrak{B}}}\right)$ is expected to be the result if BSs cannot reach an agreement. It is at least guaranteed for the BS j in the cooperative game, i.e., zero in the FEEC scheme. α_j is the normalized bargaining power. In the FEEC scheme, α_j is obtained as $\sum_{k\in\mathcal{C}_i,k\neq i}\left(\hat{l_j^i}/\hat{l_k^i}\right)$. Given this preference function, the FEEC scheme can write the RBS optimization problem as follows:

$$\begin{aligned}\mathfrak{P}^* &= \left[\mathfrak{P}_1^*\left(\hat{L}_{-i},\widehat{\Psi_1^{\mathfrak{B}}}\right)\ldots\mathfrak{P}_{i-1}^*\left(\hat{L}_{-i},\widehat{\Psi_{i-1}^{\mathfrak{B}}}\right),\mathfrak{P}_{i+1}^*\left(\hat{L}_{-i},\widehat{\Psi_{i+1}^{\mathfrak{B}}}\right)\ldots\mathfrak{P}_m^*\left(\hat{L}_{-i},\widehat{\Psi_m^{\mathfrak{B}}}\right)\right]\\ &=\arg\max_{\mathfrak{P}}\prod_{j\in\mathcal{C}_i,j\neq i}v_j\left(\hat{L}_{-i},\widehat{\Psi_j^{\mathfrak{B}}}\right),s.t.,\sum_{j\in\mathcal{C}_i,j\neq i}\mathfrak{P}_j^*\left(\hat{L}_{-i},\widehat{\Psi_j^{\mathfrak{B}}}\right)\leq\left(\Delta_e\times\left(\eta\times l_i\right)\right)\end{aligned} \quad (16)$$

Fairness-Control Game at the Second Stage

RE is generally defined as energy that comes from resources, such as sunlight, wind, tides, waves, and geothermal heat. To decrease the global greenhouse gas emissions, there are many benefits of using RE sources. For green cellular networks, RE sources can replace conventional energy grid in powering cellular BSs. It is useful not only environmental, but also economic sense while opening opportunities for new business models. Nowadays, it is of great importance to study the RE management in order to determine the potential gains and applicability scenarios (Hussein, 2013; Zhang, 2013).

Adopting RE in cellular systems affects the planning methodology and architecture of cellular platform. The green cellular network architecture consists of conventional grid-powered BSs, which are also connected by REPs. To maximize the energy efficiency, the joint design and cooperative combination of conventional grid and REPs is critical. The FEEC scheme assumes that multiple RE sources exists uniformly, and a BS is connected by only one RE source. Therefore, each BS is connected by dedicated power lines from the conventional power system and one REP. Usually, the most energy of BS is provided by the conventional grid while receiving subsidized aids from REPs. It is a quite general and real-world applicable architecture (Chia, 2014; Han, 2013; Reyhanian, 2015).

Each REP has a set of supporting BSs, called permanent cluster ($\mathcal{M}$). In a $\mathcal{M}$, each BS has its own traffic load while consuming energy differently. Due to the intermittent supply, the main goal of RE management is to fairly distribute the RE in the $\mathcal{M}$. Therefore, fairness is a new concern in the RE sharing approach. In computer science, the concept of fairness is related to the amount of delay in servicing a request that can be experienced in a shared resource environment (Etinski, 2014). In green cellular network engineering, fairness is measured whether BSs receive a fair share of RE provisioning.

In general, RE is relatively cheaper compared to electricity from traditional power-grid. Therefore, BSs would always prefer to use the RE (Zhang, 2012). The FEEC scheme is targeted to fairly distribute the limited amount of RE among different BSs. From the economic viewpoint, each BS should get the same money saving from the RE. From the viewpoint of traffic load balancing, the BS with a heavy traffic load should get the more RE. To characterize the proportional fairness of RE sharing, the FEEC scheme follow the *Jain's fairness index* ($\mathcal{F}$); it has been frequently used to measure the fairness of network operations (Dianati, 2005). According to the fundamental idea of $\mathcal{F}$, the economic fairness index ($\mathcal{F}_{e\_v}$) and the overload fairness index ($\mathcal{F}_{o\_l}$) in the $\mathcal{M}^j$ are given by

$$\mathcal{F}_{e\_v}^j = \frac{\left(\sum_{i=1}^{w} \gamma_i\left(\mathbb{A}^j\right)\right)^2}{w \times \sum_{i=1}^{w}\left(\gamma_i\left(\mathbb{A}^j\right)\right)^2} \text{ and } \mathcal{F}_{o\_l}^j = \frac{\left(\sum_{i=1}^{w} \mu_i\left(\mathbb{A}^j\right)\right)^2}{w \times \sum_{i=1}^{w}\left(\mu_i\left(\mathbb{A}^j\right)\right)^2} \tag{17}$$

where w and $\mathbb{A}^j = (\mathcal{A}_1^j, \mathcal{A}_2^j, \ldots, \mathcal{A}_w^j)$ are the number of BSs, and the RE distribution vector for each BSs in the $\mathcal{M}^j$, respectively. $\gamma_i\left(\mathbb{A}^j\right)$ is the BS i 's obtained money-saving from the RE, and $\mu_i\left(\mathbb{A}^j\right)$ is the traffic load supported from the traditional power grid. To get the proper combination of $\mathcal{F}_{e\_v}^j$ and $\mathcal{F}_{o\_l}^j$, they should be transformed into a single objective function. To provide the best compromise in the presence of different fairness indexes, a multi-objective fairness function ($M\_\mathcal{F}^{j}_{\mathcal{M}}$) for the $\mathcal{M}^{j}$ is developed based on the weighted sum method. By using dynamic joint operations, the developed $M\_\mathcal{F}^{j}_{\mathcal{M}}$ is formulated as follows.

$$M\_\mathcal{F}^{j}_{\mathcal{M}} = \left[\beta^{j} \times \mathcal{F}^{j}_{e\_v}\right] + \left[\left(1-\beta^{j}\right) \times \mathcal{F}^{j}_{o\_l}\right] \quad (18)$$

where β^{j} controls the relative weights given to different fairness indexes. Under diverse network environments, the FEEC scheme treats the β^{j} value decision problem as an on-line decision problem. When the traffic is uniformly distributed over the BSs in the $\mathcal{M}^{j}$, the FEEC scheme can put more emphasis on the economic fairness, i.e., on $\mathcal{F}^{j}_{e\_v}$. In this case, a higher value of β^{j} is more suitable. But, if traffic distributions is relatively non-uniform, due to temporal and spatial traffic fluctuations, the FEEC scheme should strongly consider the overload fairness, i.e., on $\mathcal{F}^{j}_{o\_l}$. In this case, a lower value of β^{j} is more suitable. Therefore, by considering the current traffic profiles of $\mathcal{M}^{j}$, the FEEC scheme decides the β^{j} value as follows.

$$\beta^{j} = \frac{\min\left\{\mathcal{B}^{j}_{l} \in \mathcal{M}^{j} \mid \phi^{j}_{l}\right\}}{\max\left\{\mathcal{B}^{j}_{k} \in \mathcal{M}^{j} \mid \phi^{j}_{k}\right\}} \quad (19)$$

where $\mathcal{B}^{j}_{k}$ and ϕ^{j}_{k} is the BS k and the current traffic load of BS k in the $\mathcal{M}^{j}$. Therefore, in the FEEC scheme, the value of β in each $\mathcal{M}$ is dynamically adjusted to make the system more responsive to current traffic conditions.

Summary

Research on green cellular networks is quite broad, and a number of research issues and challenges lay ahead. In particular, energy efficiency is a growing concern for cellular system operators to maintain profitability while reducing the overall environment effects. The FEEC scheme has looked into the feasibility of green cellular networks with RE sources. Employing RE is not only environment friendly, but has also other benefits with one of the notable points being the shift from energy efficiency to energy fairness. To design a novel energy control scheme, the FEEC scheme starts from a BS sleeping algorithm to maximize the energy efficiency while ensuring the fairness among BSs. Using two-level game approach, self-regarding BSs are induced to actively participate in the fair-efficient energy control mechanism.

REFERENCES

Aslam, S., & Ibnkahla, M. (2015). Optimized node classification and channel pairing scheme for RF energy harvesting based cognitive radio sensor networks. *IEEE SSD*, *2015*, 1–6.

Bao, Y., Wu, J., Zhou, S., & Niu, Z. (2015). Bayesian mechanism based inter-operator base station sharing for energy saving. *IEEE ICC*, *2015*, 49–540.

Bhowmick, A., Roy, S. D., & Kundu, S. (2015). Performance of secondary user with combined RF and non-RF based energy-harvesting in cognitive radio network. *IEEE ANTS*, *2015*, 1–3.

Chia, Y., Sun, S., & Zhang, R. (2014). Energy Cooperation in Cellular Networks with Renewable Powered Base Stations. *IEEE Transactions on Wireless Communications*, *13*(12), 6996–7010. doi:10.1109/TWC.2014.2339845

Dianati, M., Shen, X., & Naik, S. (2005). A New Fairness Index for Radio Resource Allocation in Wireless Networks. In *Proc. of IEEE WCNC*, (pp. 712-715). doi:10.1109/WCNC.2005.1424595

Etinski, M., & Schulke, A. (2014). Fair sharing of RES among multiple users. *IEEE ISGT*, *2014*, 1–5.

Feng, W., & Jaafar, M. (2009). Energy Efficiency in Ad-hoc Wireless Networks with Two Realistic Physical Layer Models.*IEEE Third International Conference on Next Generation Mobile Applications, Services and Technologies*, (pp. 401-406). doi:10.1109/NGMAST.2009.30

Han, T., & Ansari, N. (2013). On Optimizing Green Energy Utilization for Cellular Networks with Hybrid Energy Supplies. *IEEE Transactions on Wireless Communications*, *12*(8), 3872–3882. doi:10.1109/TCOMM.2013.051313.121249

Hoang, D., Niyato, D., Wang, P., & Kim, D. (2014). Opportunistic Channel Access and RF Energy Harvesting in Cognitive Radio Networks. *IEEE Journal on Selected Areas in Communications*, *32*(11), 2039–2052. doi:10.1109/JSAC.2014.141108

Hussein, A., Nuaymi, L., & Pelov, X. (2013). Renewable energy in cellular networks: A survey. *IEEE GreenCom, 2013*, 1–7.

Khansalee, E., Zhao, Y., Leelarasmee, E., & Nuanyai, K. (2014). A dual-band rectifier for RF energy harvesting systems. *IEEE ECTI-CON, 2014*, 1–4.

Kim, S. (2014). *Game Theory Applications in Network Design*. IGI Global. doi:10.4018/978-1-4666-6050-2

Lim, T., Lee, N., & Poh, B. (2013). Feasibility study on ambient RF energy harvesting for wireless sensor network. *IEEE IMWS-BIO, 2013*, 1–3.

Lu, X., Wang, P., Niyato, D., & Hossain, E. (2014). Dynamic spectrum access in cognitive radio networks with RF energy harvesting. *IEEE Wireless Communications*, 102-110.

Lu, X., Wang, P., Niyato, D., Kim, D., & Han, Z. (2015). Wireless Networks With RF Energy Harvesting: A Contemporary Survey. *IEEE Communications Surveys and Tutorials*, *17*(2), 757–789. doi:10.1109/COMST.2014.2368999

Niyato, D., Wang, P., & Kim, D. (2014). Admission control policy for wireless networks with RF energy transfer. *IEEE ICC, 2014*, 1118–1123.

Park, S., Heo, J., Kim, B., Chung, W., Wang, H., & Hong, D. (2012). Optimal mode selection for cognitive radio sensor networks with RF energy harvesting. *IEEE PIMRC, 2012*, 2155–2159.

Reyhanian, N., Vahid, S., Maham, B., & Yuen, C. (2015). Renewable energy distribution in cooperative cellular networks with energy harvesting. *IEEE PIMRC, 2015*, 1617–1621.

Wang, B., Liu, K. J. R., & Clancy, T. C. (2010). Evolutionary cooperative spectrum sensing game: How to collaborate. *IEEE Transactions on Communications*, *58*(3), 890–900. doi:10.1109/TCOMM.2010.03.090084

Wu, J., Zhang, Y., Zukerman, M., & Yung, E. (2015). Energy-Efficient Base-Stations Sleep-Mode Techniques in Green Cellular Networks: A Survey. *IEEE Communications Surveys and Tutorials*, *17*(2), 803–826. doi:10.1109/COMST.2015.2403395

Wu, Y., & Song, W. (2014). Cooperative Resource Sharing and Pricing for Proactive Dynamic Spectrum Access via Nash Bargaining Solution. *IEEE Transactions on Parallel and Distributed Systems*, *25*(11), 2804–2817. doi:10.1109/TPDS.2013.285

Zhang, C., Wu, W., Huang, H., & Yu, H. (2012). Fair energy resource allocation by minority game algorithm for smart buildings. *IEEE DATE, 2012*, 63–68.

Zhang, H., Cai, J., & Li, X. (2013). Energy-efficient base station control with dynamic clustering in cellular network. *IEEE CHINACOM, 2013*, 384–388.

Zheng, J., Cai, Y., Chen, X., Li, R., & Zhang, H. (2015). Optimal Base Station Sleeping in Green Cellular Networks: A Distributed Cooperative Framework Based on Game Theory. *IEEE Transactions on Wireless Communications*, *14*(8), 4391–4406. doi:10.1109/TWC.2015.2420233

KEY TERMS AND DEFINITIONS

Gaussian Noise: Statistical noise having a probability density function equal to that of the normal distribution, which is also known as the Gaussian distribution.

Green Networking: Energy efficiency has come into focus for information and communication technology due to the global movement of energy saving related to the global climate change. Green networking is the practice of selecting energy-efficient networking technologies and products, and minimizing resource use whenever possible.

Pareto Optimality: A state of allocation of resources in which it is impossible to make any one individual better off without making at least one individual worse off. The concept has applications in academic fields such as economics, engineering, and the life sciences.

Radio Frequency: Any of the electromagnetic wave frequencies that lie in the range extending from around 3 kHz to 300 GHz, which include those frequencies used for communications or radar signals. It usually refers to electrical rather than mechanical oscillations.

TDMA: Time division multiple access (TDMA) is a channel access method for shared medium networks. It allows several users to share the same frequency channel by dividing the signal into different time slots. The users transmit in rapid succession, one after the other, each using its own time slot. This allows multiple stations to share the same transmission medium while using only a part of its channel capacity

Chapter 6
Developing IoT Applications for Future Networks

ABSTRACT

Applications in the IoT domain need to manage and integrate huge amounts of heterogeneous devices. Usually these devices are treated as external dependencies residing at the edge of the infrastructure mainly transmitting sensed data or reacting to their environment. Recently, these devices will fuel the evolution of the IoT as they feed sensor data to the Internet at a societal scale. Leveraging volunteers and their mobiles as a sensing data collection outlet is known as Mobile Crowd Sensing (MCS) and poses interesting challenges, with particular regard to the management of sensing resource contributors, dealing with their subscription, random and unpredictable join and leave, and node churn. In addition, with the advent of new wireless technologies, it is expected that the use of Machine-Type Communication (MTC) will significantly increase in next generation IoT. MTC has broad application prospects and market potential. In this chapter, we explore new IoT applications for future IoT paradigms.

DOI: 10.4018/978-1-5225-1952-2.ch006

THE MACHINE TYPE COMMUNICATION CONTROL (MTCC) SCHEME

For the next generation wireless network, Machine Type Communication (MTC) is gaining an enormous interest as a new communication paradigm. MTC is expected to become a cost-effective solution for improving the wireless communication performance. In MTC, one of the most critical issues is to support data transfers among devices without human interactions. The *Machine Type Communication Control* (MTCC) scheme is a new MTC control scheme for the future network infrastructure. To effectively support a large number of MTC devices, the MTCC scheme investigates a dual-level interaction mechanism by employing a timed strategy game model.

Development Motivation

In the past decade, mobile data traffic services have been experiencing a phenomenal rise. This ever-increasing data traffic puts significant pressure on the development of a new state-of-the-art communication method. As a new wireless communication paradigm, Machine Type Communication (MTC) is gaining a tremendous attention among mobile network operators and equipment vendors. MTC can support various automated operations without or with minimal human interactions (Niyato, 2014). In particular, it aims to provide ubiquitous connectivity to a variety of network devices in smart metering, telematics, automobile, smart cities, etc. With the different traffic services, MTC can be used in almost everywhere in our everyday life (Kouzayha, 2013; Niyato, 2014).

Due to the unique features of MTC, the MTC method poses several challenges to network operators. First, it is expected to significantly increase the total amount of traffic services. This will definitely cause intense competitions for the wireless spectrum resource. Second, quality of service (QoS) provisioning should be considered seriously. It is essential to ensure different traffic characteristics. Third, system efficiency with the limited network resource is another prominent issue for the MTC management. To satisfy these conflicting requirements, we need a new concept for effective system-wide solutions (Kouzayha, 2013).

During MTC operations, adaptive power control algorithm is a key technique to deal with above challenges. In particular, transmission power control is essential to satisfy QoS requirements while reducing interference and energy consumption (Chaves, 2013). In the wireless spectrum management, spectrum sharing and QoS ensuring are translated into requirements on the quality of received signal, such as the obtained *Signal-to-Interference-plus-Noise Ratio* (SINR) at the receiver. Therefore, the task of power control is to dynamically adjust the transmission power to the minimum value so that a desired SINR level can be attained at the receiver. Usually, traditional power control algorithms have assumed the perfect knowledge of link quality information. However, this information is subject to errors due to aspects like power control loop delay and measurement uncertainties. Therefore, traditional approaches are impractical methods (Chaves, 2013).

To design a novel and realistic power control algorithm for MTC systems, it is necessary to study a strategic decision making process in each MTC device. Under widely dynamic MTC environments, MTC devices can be assumed as intelligent rational decision-makers, and they select a best-response strategy to maximize their expected payoff in a distributed fashion. This situation is well-suited for the game theory.

The main goal of the MTCC scheme is to maximize system performance while ensuring QoS guarantees. In dynamically changing MTC environments, each individual MTC device can constantly adapt its power level in a non-cooperative game manner. Based on the received power levels, the system operator allocates adaptively the total spectrum resource in a cooperative manner. To effectively handle this reciprocal interaction mechanism over time, the MTCC scheme adopts a timed strategic game model. With the incomplete information, the timed strategic approach can relax the traditional assumption in game theory that all information is completely known; this is the main advantage of the MTCC scheme.

MTC System Architecture and Timed Strategic Game Model

The MTCC scheme considers a MTC-enabled cellular network infrastructure while including multiple MTC devices. If all MTC devices send transmission requests to a Base Station (BS) directly, intense traffic congestions occur. It may lead to the inefficient utilization of spectrum resource. To increase the efficiency of system resources, a new network element, called as MTC gateway, is deployed. MTC gateway can collect data from a group of MTC devices and forward the data to the BS over cellular uplinks. Therefore, the MTC gateway is responsible for managing power consumptions and providing suitable communication paths between MTC devices and network system (Zhang, 2016).

In the MTCC scheme, the BS, multiple MTC gateways and MTC devices are collocated for data transmissions. Each individual MTC devices constantly adapt their power levels while satisfying required QoS constraints. To increase the resource utilization, the BS adaptively allocates the wireless spectrum into MTC gateways based on the timed bargaining process. Therefore, the whole available resource is divided to properly balance the traffic loads in MTC gateways. This approach not only enhances the energy efficiency, but also enables the separate and flexible QoS control process to each type of traffic services. From a MTC device level view, the main goal is to maximize individual payoff. From the system level view, the major concern is to upgrade the resource efficiency while maximizing system performance. Through coordinating the transmit power and spectrum sharing mechanisms, the MTCC scheme can strike an appropriate performance balance.

In the game theory, strategic decisions should be made to optimize the objectives at a given time instant. However, on a long time-scale, these decisions are not necessarily optimal to evaluate the long-term system performance. Therefore, timed strategic decisions are needed to get insights into the long-term effective decisions under diversified MTC network situations. The MTCC scheme introduces a new timed strategic game model ($\mathbb{G}$) for MTC systems. $\mathbb{G}$ is a tuple

$$\left(BS, \mathbb{M}, \mathcal{R}, \mathfrak{A}, \mathbb{N}, \left(\boldsymbol{S}_i\right)_{i\in\mathbb{N}}, \left(U_i\right)_{i\in N}, T\right)$$

at each time period t of gameplay.

- BS is the base station and $\mathbb{M} = \left\{\mathcal{G}_1, \ldots \mathcal{G}_c \ldots \mathcal{G}_m\right\}$ is the finite set of MTC gateways where $\mathcal{G}_{c,1\leq c\leq m}$.
- $\mathcal{R}$ is the total amount of available spectrum resource in the MTC-enabled cellular network.
- $\mathfrak{A}$ is the vector of spectrum allocation for each MTC gateways; $\mathfrak{A} = \left\{\mathcal{A}_{\mathcal{G}_1} \ldots \mathcal{A}_{\mathcal{G}_m}\right\}$ where $\sum_{z\in\mathbb{M}} \mathcal{A}_z \leq \mathcal{R}$.
- $\mathbb{N}$ is the finite set of MTC devices $\mathbb{N} = \left\{1, \ldots i \ldots n\right\}$ where $i_{1\leq i\leq n}$ represents the ith MTC device.
- $\boldsymbol{S}_i$ is the set of strategies of the MTC device i. $\boldsymbol{S}_i$ is defined as all possible power levels for the device i.
- The U_i is the payoff received by the MTC device i. Traditionally, the payoff is determined as the obtained outcome minus the cost to obtain that outcome.

- The T is a time period. The $\mathbb{G}$ is repeated $t \in T < \infty$ time periods with competitive and cooperative manner.

To effectively operate the interactions of BS, MTC gateways and MTC devices, the timed strategic game has some attractive characteristics. Most of all, the most important feature is to adopt a timed interactive feedback mechanism. Therefore, the MTC system can adaptively respond to the current network situations over a game processing time. Initially, self-regarding MTC devices dynamically adjusts their power levels to maximize their payoffs. In this stage, the distributed power control procedure is formulated as a non-cooperative game model. At the same time, the BS adaptively distributes the spectrum resource to multiple MTC gateways based on the current traffic situation. In this stage, the central resource distribution procedure is formulated as a cooperative bargaining game model. Through adaptively making timed strategic decisions, these two different game approaches are sophisticatedly combined to approximate a well-balanced system performance.

Timed Strategic Power Control Algorithm

Due to the interference interactions in wireless communications, power decisions made by MTC devices will affect the performance of the other MTC devices. Therefore, the main goal of power control problem in MTC systems is to decide how the wireless spectrum is shared among different devices while maximizing total network performance. In the game model of the MTCC scheme, individual MTC devices are defined as game players. Each player attempts to maximize his throughput by adaptively selecting the transmission power (p_l). The set of power strategies ($\boldsymbol{S}_i$) of player i is defined as below.

$$\boldsymbol{S}_i = \left\{ \Pi_{i \in \mathbb{N}} p\_l_l^i \mid p\_l_l^i \in \left[p\_l_{\min}^i, p\_l_{\max}^i \right] \right\} \tag{1}$$

where $\mathbb{N}$ is the number of players and $p\_l_l^i$ is the lth transmission power level of player i. The $p\_l_{\min}^i, p\_l_{\max}^i$ are the pre-defined minimum and maximum power levels, respectively. The player i selects a single strategy from the $\boldsymbol{S}_i$, and estimates the expected throughput. By considering the tradeoff of bit error rate, interference and energy consumption, the throughput of MTC i (T_i) is defined as the ratio of the throughput to transmit power.

$$\mathcal{T}_i\left(\mathbb{P}\right)=\left(\psi_i\times\log_2\left(1+\left\{\frac{\left(\gamma_i\left(\mathbb{P}\right)\right)}{\Omega}\right\}\right)\right)\Big/p\_l_l^i \quad s.t.,\gamma_i\left(\mathbb{P}\right)=\frac{\left(p\_l_l^i\times h_{ii}\right)}{\left[\sigma_i+\sum_{j\neq i}\left(p\_l_g^j\times h_{ji}\right)\right]} \tag{2}$$

where $\mathbb{P}=(p\_l_{(\cdot)}^1,\ldots p\_l_l^i\ldots p\_l_{(\cdot)}^n)$ is the transmit power vector for all players and ψ_i is the allocated wireless spectrum for the player i. Ω (Ω ≥ 1) is the gap between uncoded M-QAM and the capacity, minus the coding gain. σ_i is the background noise within the player's spectrum, h_{ji} is the path gain from the transmitter of player j to the receiver of player i (Yang, 2009).

Traditionally, heterogeneous traffic services can be categorized into two classes according to the required QoS: *class I* (real-time) traffic services and *class II* (non real-time) traffic services. *Class I* data services are highly delay sensitive, and strict deadlines are applied. However, some flexible data services are called as *class II* traffic services; they are rather tolerant of delays. Different multimedia services have different QoS requirements (Yang, 2009). To support the miscellaneous QoS provisioning, the MTCC scheme defines a QoS function ($Q\_F_k^t$) for the traffic service *k* ($\mathcal{D}_k$) at time t. Based on the different data characteristics, it is formulated as follows;

$$Q\_F_k^t\left(\mathcal{D}_k^{max},\mathcal{D}_k^{min},\mathcal{D}_k^t\right)=\frac{x}{x+e^{-x}}\times\left[x\times\log\left(x+1\right)\right]$$

s.t.,

$$x=\begin{cases}\min\left\{1,\dfrac{\mathcal{D}_k^t}{\mathcal{D}_k^{\max}}\right\}, & if\,\mathcal{D}_k\ is\ class\ I\ data\\ \dfrac{\mathcal{D}_k^t}{\mathcal{D}_k^{\max}\times\left(\mathcal{D}_k^t-\mathcal{D}_k^{\min}\right)}\times\max\left\{0,\left(\mathcal{D}_k^t-\mathcal{D}_k^{\min}\right)\right\}, & otherwise\end{cases} \tag{3}$$

where $\mathcal{D}_k^{\max}$ and $\mathcal{D}_k^{\min}$ are the pre-defined maximum and minimum data transmission rate for the traffic service *k* ($\mathcal{D}_k$), respectively. $\mathcal{D}_k^t$ is the data transmission rate at time t. To specify the nature of mapping actions and time to real numbers, the utility function is continuous and timed-payoff function with a time parameter. Each individual player attempts to maximize his payoff by adaptively selecting power level. By combining (2) with (3), the utility function for the player i at time t ($U_i^t(\cdot)$) can be defined as follows;

$$U_i^t\left(p\_l^i(t), \mathcal{D}_i^t, \psi_i^t\right) = \int_0^t \max_{p\_l^i(t) \in \mathbf{S}_i} \left(\ln\left(\mathcal{T}_i\left(\mathbb{P}^t\right)\right) + \ln\left(Q\_F_k^t\left(\mathcal{D}_i^{\max}, \mathcal{D}_i^{\min}, \mathcal{D}_i^t\right)\right)\right) dt \tag{4}$$

where $p\_l^i(t)$, ψ_i^t and $\mathbb{P}^t$ are the player i's transmit power level, allocated spectrum, and the $\mathbb{P}$ vector at time t, respectively. By taking into account the current time when an action is taken and how long an action lasts, players have the opportunity to see how the payoff stemming from the actions taken by them unfolds over time. It is a repeated timed strategic game model (Sarcia, 2013).

The timed strategic game is the setting to explain not only what to do to win the game as usual, but also how the payoff function will progress over time giving more insights on the evolution of the game for prediction purposes. Therefore, the setting of developed timed strategic game is the same as the usual one apart from the one-shot game that is defined over the variable time t. To represent the power control problem at each time, let $\eta_j^i(t-1)$ is the player i's propensity for the jth power level, i.e., $p\_l_j^i$, at time t-1. When a player selects the $p\_l_j^i$ at time t-1, $\eta_{j \in \mathbf{S}_i}^i(t)$ is dynamically estimated as follows; it is obtained in a distributed manner for each individual players.

$$\eta_j^i(t) = \eta_j^i(t-1) \times \left\{ 1 + \frac{U_i^{t-1}\left(p_l^i(t-1), \mathcal{D}_i^{t-1}, \psi_i^{t-1}\right) - \left(\left(1/(t-1)\right) \times \sum_{k=1}^{t-1}\left(\int_{k-1}^{k}\left[U_i^k\right] dk\right)\right)}{\left(1/(t-1)\right) \times \sum_{k=1}^{t-1}\left(\int_{k-1}^{k}\left[U_i^k\right] dk\right)} \right\} \tag{5}$$

According to (5), the strategy j 's selection probability of player i ($\mathcal{P}_j^i$) can be captured based on the timed empirical outcomes. Formally, $\mathcal{P}_j^i$ at the time t $\left(\mathcal{P}_j^i\left(t\right)\right)$ is calculated using the timed weighted average of propensities;

$$\mathcal{P}_j^i\left(t\right) = \sum_{k=1}^{t}\left(\left(1-\mu\right)^{t-k} \times \eta_j^k\right) / \left(\sum_{l \in \boldsymbol{S}_i}\sum_{k=1}^{t}\left(\left(1-\mu\right)^{t-k} \times \eta_l^k\right)\right) \tag{6}$$

where μ is a control factor, which can control the weight for each timed strategies. In a distributed online fashion, individual players dynamically estimates each $\mathcal{P}_{j \in \boldsymbol{S}_i}^{i,1 \le i \le n}\left(\cdot\right)$ values for each strategies.

Spectrum Sharing Algorithm among MTC Gateways

According to the current traffic load, the BS dynamically allocate the spectrum resource to multiple MTC gateways. In recent years, cooperative approaches derived from game theory have been widely used for efficient resource allocation problems. Because of many appealing properties, the most popular approach is the *Nash Bargaining Solution* (NBS). NBS can fulfill the six axioms - *Individual rationality*, *feasibility*, *Symmetry*, *Pareto optimality*, *Independence of irrelevant alternatives* and *Invariance with respect to utility transformations*, and provide a unique and fair-efficient bargaining solution (Kim, 2014). However, traditional NBS assumes the static situation while knowing all information completely. In a real world situation, the information about MTC system is dynamically changing. Therefore, classical NBS model cannot be directly applied for real MTC network operations.

The MTCC scheme develops a timed NBS (*T_NBS*) model based on the reinforcement bargaining power, which represents dynamically the current traffic environment of each MTC gateway. According to the timed bargaining process, the BS adaptively allocates the spectrum resource in the iterative online manner. When the traffic condition in the MTC system is changed over time, the timed bargaining process can provide a good trade-off between optimized network performance and the complexity of practical implementation. By considering the payoffs of individual MTC devices, the utility function of MTC gateway $\mathcal{G}_c$ ($\varrho^{\mathcal{G}_c}\left(\cdot,t\right)$) at the time t is defined as follows;

$$\varrho^{\mathcal{G}_c}\left(\mathcal{A}_{\mathcal{G}_c}^t, t\right) = \sum_{i \in \mathcal{G}_c} U_i^t\left(p\_l^i\left(t\right), \mathcal{D}_i^t, \psi_i^t\right)$$

s.t.,

$$\mathcal{G}_c \in \mathbb{M}, \mathcal{A}_{\mathcal{G}_c}^t \mapsto \left\{\psi_i^t \mid i \in \mathcal{G}_c\right\} \text{and } \mathcal{A}_{\mathcal{G}_c}^t \leq \sum_{i \in \mathcal{G}_c} \psi_i \tag{7}$$

where $\mathcal{A}_{\mathcal{G}_c}^t$ is the amount of assigned spectrum for the $\mathcal{G}_c$ at the time t. It is distributed for individual traffic services, which are covered by $\mathcal{G}_c$. Based on the all $\varrho^{\mathcal{G}_c}\left(\mathcal{A}_{\mathcal{G}_c}^t, t\right)$ values where $\mathcal{G}_c \in \mathbb{M}$, *T_NBS* is given by;

$$T\_NBS = \left[\mathcal{A}_{\mathcal{G}_1}^t \ldots \mathcal{A}_{\mathcal{G}_c}^t \ldots \mathcal{A}_{\mathcal{G}_m}^t\right] = \max_{\mathcal{A}_{\mathcal{G}_c}^t} \int_0^t \left(\sum_{\mathcal{G}_c \in \mathbb{M}} \left(\gamma^{\mathcal{G}_c}(t) \times \ln\left(\varrho^{\mathcal{G}_c}\left(\mathcal{A}_{\mathcal{G}_c}^t, t\right)\right)\right)\right) dt$$

s.t.,

$$\sum_{\mathcal{G}_c \in \mathbb{M}} \mathcal{A}_{\mathcal{G}_c}^t \leq \mathcal{R} \text{ and } \gamma^{\mathcal{G}_c}(t) = \frac{\mathcal{T}_I^{\mathcal{G}_c}(t)}{\mathcal{T}_I^{\mathcal{G}_c}(t) + \mathcal{T}_{II}^{\mathcal{G}_c}(t)} \tag{8}$$

where $\gamma^{\mathcal{G}_c}(t)$ is the bargaining power of $\mathcal{G}_c$ at time t. In the $\mathcal{G}_c$, $\mathcal{T}_I^{\mathcal{G}_c}(t)$ (*or* $\mathcal{T}_{II}^{\mathcal{G}_c}(t)$) is the requested spectrum amount of *class I* (*or class II*) traffic services at time t. Usually, the bargaining solution is strongly dependent on the bargaining powers. If different bargaining powers are used, the player with a higher bargaining power obtains a higher profit than the other players. In the MTCC scheme, by considering the QoS preference, *class I* traffic services are strictly preferred over the *class II* traffic services. During the step-by-step iteration, a fair-efficient *T_NBS* can be obtained while providing a suitable balance between the desired QoS and high spectrum utilization.

Summary

During the last decade, several techniques have been developed for timed planning problems. These techniques were based on constraint programming and gave good performances in practice. However, a major difficulty was related to the fast system condition fluctuations and non-scalability of control algorithms. Recently, timed strategic game model has been introduced in order to represent timed systems, and extensively studied both from a theoretical and applied viewpoints. The MTCC scheme is proposed as a novel MTC control algorithm based on the timed strategic game model. The main goal of the MTCC scheme is to maximize system performance while ensuring service QoS. To satisfy this conflicting goals, the MTCC scheme develops a timed feedback mechanism based on the step-by-step interactive process.

THE INTERACTIVE MOBILE CROWD SENSING (IMCS) SCHEME

With the fast increasing popularity of mobile services, ubiquitous mobile devices with enhanced sensing capabilities collect and share local information towards a common goal. The recent Mobile Crowd Sensing (MCS) paradigm enables a broad range of mobile applications and undoubtedly revolutionizes many sectors of our life. A critical challenge for the MCS paradigm is to induce mobile devices to be workers providing sensing services. The *Interactive Mobile Crowd Sensing* (IMCS) scheme examines the problem of sensing task assignment to maximize the overall performance in MCS system while ensuring reciprocal advantages among mobile devices. Based on the overlapping coalition game model, the IMCS scheme can effectively decomposes the complex optimization problem and obtains an effective solution using the interactive learning process.

Development Motivation

With the development of IoT and embedded technology, the remote intelligent monitoring system will be applied in more comprehensive scopes. Therefore, ubiquity of internet-connected portable devices is enabling a new class of applications to perform sensing tasks in the real world. Among mobile devices, smartphones have evolved as key electronic devices for communications, computing, and entertainment, and have become an important part of people's daily lives. Most of current mobile phones are equipped with a rich set of embedded sensors, which can also be connected to a mobile phone via its Bluetooth interface. These sensors can enable attractive sensing applications in various domains such as environmental monitoring, social network, healthcare, transportation, and safety (An, 2015; Salem, 2013; Sheng, 2014).

Mobile Crowd Sensing (MCS) refers to the technology that uses mobile devices, i.e., smartphones, to collect and analyze the information of people and surrounding environments. Based on this information, we can analyze statistical characteristics of group behaviors, reveal hidden information of social activity patterns, and finally provides useful information and services to end users. By involving anyone in the process of sensing, MCS greatly extends the service of IoT and builds a new generation of intelligent networks that interconnect things-things, things-people and people-people. Therefore, to provide a new way of perceiving the world, MCS has a wide range of potential applications (An, 2015).

To effectively operate the MCS system, we collect sensing data from multiple smartphones to maximize the utility of sensed information; different smartphones are intrinsically different in terms of the quality of their embedded sensors. Generally, it is desirable to obtain more sensing data from high quality sensors. However, performing a sensing task consumes precious resources on smartphones, such as energy, computing, and cellular bandwidth. Therefore, one of the central problems for MCS systems is allocation of sensing-workload among smartphones, which has a great impact on the overall MCS system performance. Furthermore, the MCS platform publicizes multiple tasks for the smartphones to choose, in order to support various smartphone applications. It makes the sensing-workload allocation problem more challenging (Di, 2013; Li, 2014).

In the early research of MCS system, smartphones are assumed as volunteers for the MCS system. However, in practice, smartphone users are always selfish individuals that will not contribute their resources without getting paid. Therefore, the traditional volunteer models may not be suitable for the real-world MCS operation (Di, 2013). This situation can be seen a game theory problem. Coalitional games are very influential game models for the multi-agent system research community due to their ability to capture the cooperative behaviors among agents (Zhan, 2012). With respect to this activity, there are many approaches that try to generate the optimal coalition structure which maximizes the payoffs of all agents. However, these approaches are all based on the assumption that each agent must belong to one coalition. This means that even if an agent has excessive resources to participate other coalitions, it is still not allowed to take advantage of the remaining resources (Zhan, 2012). In 2010, Chalkiadakis et al. introduced the new concept of Overlapping Coalition Formation (OCF) game (Chalkiadakis, 2010). In OCF games, individual agents can cooperate with each other to form coalitions, and the coalitions can be overlapping.

Typically, smartphones consume their own resources to accomplish the sensing task. Therefore, the MCS system should purchase sensing services from smartphones to compensate their resource consumptions. Without compensation of a plenty of smartphones, the MCS system is not able to collect enough information to accomplish its task. The IMCS scheme focuses on the multi-workload allocation problem among smartphones. To attract sufficient participations, paying price influences the willingness of smartphones to serve. However, it is not easy to characterize behaviors of smartphone users who may decide their actions to serve. In particular, as the number of smartphones can be huge, it is difficult to apply conventional optimization methods. They pose a heavy computation burden and implementation overheads.

To address this issue, the IMCS scheme would employ the OCF game methodology to develop a novel multi-workload allocation algorithm for MCS systems. This approach decomposes the complex optimization problem into several sub-problems, and each sub-problem is solved through an interactive procedure imitating the negotiation process. It is practical and suitable for real implementation. Based on the iterative feedback mechanism, the IMCS scheme dynamically adjusts the price for each sensing task, and smartphone users individually respond to such price setting in order to optimize their payoffs. Under the real-world MCS environments, the central server and smartphone users are mutually dependent on each other to maximize their profits while flexibly adapting the current system situations.

Overlapping Coalition Formation Game Model

To model cooperative games with overlapping coalitions, it is assumed that players possess a certain amount of resources which they can distribute among the coalitions they join. If overlapping coalitions are allowed, players selectively participate some coalitions, and players' contribution to a coalition is given by the fraction of their resources that they allocate to it. In the traditional non-overlapping coalition formation game, a coalition is a subset of players, and a game is defined by its characteristic function $v : 2^N \rightarrow \mathbb{R}$ with player set $N = \{1,\ldots,n\}$, representing the maximum total payoff that a coalition can get (Chalkiadakis, 2010).

In the OCF game, a coalition *k* is given by a vector $\boldsymbol{r}^k = \left(r_1^k,\ldots,r_n^k\right)$, where r_i^k is the fraction of agent i 's resources contributed to the coalition *k*; $r_i^k = 0$ means that the player i is not a member of the coalition *k*. The *support* of coalition *k*, denoted by $\text{supp}\left(\boldsymbol{r}^k\right)$, is given by

$$\text{supp}\left(\boldsymbol{r}^k\right) = \left\{i \in N \mid r_i^k \neq 0\right\}.$$

The OCF game is given by a characteristic function

$$v(k):\left[0, r_{i,1\le i\le n}^{k}\right]^{n} \rightarrow \mathfrak{R},$$

where $v(0^{n})=0$ and $\mathfrak{R}$ represents the set of real numbers. Function $v(k)$ is *monotone*, and maps each contribution r_i^k in $\boldsymbol{r}^k$ to the corresponding payoff (Chalkiadakis, 2010).

The IMCS scheme develops a new OCF game model ($\mathbb{G}$) for MCS system. To model strategic MCS situations involving interactive process, the IMCS scheme assumes that players seek to choose their strategy based on the reciprocal relationship. In the game model, each coalition represents individual sensing task in the MCS system.

Definition 1: OCF game model constitutes a 8-tuple $\mathbb{G} = (\boldsymbol{N} \cup \{0\}, m, \boldsymbol{r}^{(\cdot)}, v(\cdot), \mathcal{T}, \mathbb{A}, \boldsymbol{S}_{i,1\le i\le n}, \boldsymbol{U}_{i,1\le i\le n})$, where

1. $\boldsymbol{N} \cup \{0\}$ is a set of game players; $\boldsymbol{N} = \{1,\ldots,n\}$ is the set of smartphone users and 0 represents the central server of MCS system,
2. m is the number of sensing tasks in the MCS system,
3. $\boldsymbol{r}^{(\cdot)}$ is a vector to represent the contributed resources from players in $\boldsymbol{N}$,
4. $v(\cdot)$ represents a satisfaction level of each task. It is a *monotone* function to evaluate each task payoff. Therefore, v function satisfies $v(\boldsymbol{r}^k) \ge v(\boldsymbol{r}'^k)$ for any $\boldsymbol{r}^k$, $\boldsymbol{r}'^k$ such that $r_i^k \ge r_i'^k$ for all $i \in \boldsymbol{N}$.
5. $\mathcal{T} = \{\mathfrak{T}^1,\ldots,\mathfrak{T}^k,\ldots,\mathfrak{T}^m\}$ is a set of each task's thresholds, $\mathfrak{T} > 0$. If $\boldsymbol{r}^k < \mathfrak{T}^k$, $v(\boldsymbol{r}^k) = 0$. This means that players must allocate resource at least the $\mathfrak{T}^k$ amount to complete the task k,
6. $\mathbb{A} = \{\mathcal{A}_1,\ldots,\mathcal{A}_i,\ldots,\mathcal{A}_n\}$ is a set of available resources for players in $\boldsymbol{N}$. $\mathcal{A}_i$ represents the total resource amount of the player $i \in \boldsymbol{N}$. For the sake of simplicity, the IMCS scheme assumes one type of resource, e.g., sensing capacity, that is needed for all tasks,
7. $\boldsymbol{S}_i = \{s_i^1,\ldots,s_i^k,\ldots,s_i^m\}$ is a nonempty finite set of all pure strategies of the player $i \in \boldsymbol{N} \cup \{0\}$. In particular, if $i = 0$, s_i^k represents the central server's price strategy for the task k. Otherwise, if $i \in \boldsymbol{N}$, s_i^k represents the smartphone i's contribution to the task k.

8. $\boldsymbol{U}_i = \left\{u_i^1,\ldots,u_i^k,\ldots,u_i^m\right\}$ is the utility set of the player $i \in \boldsymbol{N} \cup \left\{0\right\}$. Therefore, u_i^k represents the player *i*'s payoff for the task *k*. If $i =$ 0, $\boldsymbol{U}_0$ represents a satisfaction level of the central server: ($s_0^1 \times s_0^2 \times \cdots \times s_0^m$) $\rightarrow \boldsymbol{U}_0 = \sum_{k=1}^{m} v\left(\boldsymbol{r}^k\right)$. Otherwise, if $i \in \boldsymbol{N}$, $\boldsymbol{U}_i$ represents a satisfaction level of player *i*, and it is decided according to the set of player *i*'s strategies: ($s_i^1 \times s_i^2 \times \cdots \times s_i^m$) $\rightarrow \boldsymbol{U}_i = \sum_{k=1}^{m} u_i^k$.

In the IMCS scheme, the OCF game model describes a scenario where players can split their resources to work different m tasks. Each task has its own resource requirement $\mathfrak{T}$ and a utility function $v\left(\cdot\right)$. If the $\mathrm{supp}\left(\boldsymbol{r}^k\right)$, which is the total sum of contribution resources from players that work on the task *k*, is higher than the $\mathfrak{T}^k$, the task *k* has sufficient resources to be completed, and the central server can obtain the outcome of $v\left(\boldsymbol{r}^k\right)$. Otherwise, the payoff from the task *k* is 0. Without loss of generality, the IMCS scheme assumes that each player chooses independently to work on each individual task while not preventing another players from choosing any tasks as well. For real-world MCS operations, this assumption is generally holds.

Interactive Mobile Crowd Sensing Algorithm

The IMCS scheme considers a multitask-oriented central server, and n smartphone users randomly spread over the MCS region. Each smartphone is embedded with various sensors, and different sensing-works are assumed to be independent of each other, both temporarily and spatially. The central server publicizes m sensing tasks, and gathers the sensing information from the sensory data contributors, i.e., smartphones, in the MCS region. The total sensing resource of each smartphone is limited to $\mathcal{A}$, and smartphone users can distribute their resources freely among m tasks. Each task is associated with a set of thresholds ($\mathcal{T}$). Therefore, for each task completion, smartphone users need to upload at least the minimum amount $\mathfrak{T}$ from their devices.

To evaluate each task's sensing performance, the central server has its own utility function, which is a function of the participating smartphones and their corresponding contributions. To recruit smartphones for collecting sensory data, a proper payment mechanism should be employed to model the interactions between smartphone users and the central server. By using the reciprocal relationship between benefit and cost, the central server's payoff corresponds to the received outcome minus the incurred cost. In the central server, the utility function for the task *k* is defined as follows;

$$\boldsymbol{U}_0\left(\boldsymbol{s}_0^{k,1\le k\le m}\right) = \sum_{k=1}^{m} v\left(\boldsymbol{r}^k\right) = \sum_{k=1}^{m}\left(\zeta^k - \mathcal{C}_0\left(\boldsymbol{r}^k\right)\right)$$

s.t.,

$$\boldsymbol{r}^k = \sum_{l=1}^{n} r_l^k \text{ and } \mathcal{C}_0\left(\boldsymbol{r}^k\right) = \mathcal{P}^k \times \boldsymbol{r}^k \tag{9}$$

where $\mathcal{P}^k$ and ζ^k are the *unit_price* for resource and the obtained outcome from the task k completion, respectively. If $\boldsymbol{r}^k < \mathfrak{T}^k$, the ζ^k value is 0. According to the equation (9), the $\boldsymbol{U}_0(\cdot)$ function has a nice interpretation: the net gain of central server's utility decreases proportionately by the sensing payment. In contrast to the central server, the net gain of smartphones increases proportionately by the sensing payment. Therefore, the individual utility function for the player $i \in \boldsymbol{N}$ ($\boldsymbol{U}_i\left(\boldsymbol{S}_i\right)$) is defined as follows;

$$\boldsymbol{U}_i\left(\boldsymbol{S}_i\right) = \sum_{k=1}^{m} u_i^k = \sum_{k=1}^{m}\left(\left\{\frac{\mathcal{P}^k}{1+\exp\left(-\zeta^k \times \dfrac{s_i^k}{\mathcal{A}_i}\right)}\right\} - \mathcal{C}_i\left(k, s_i^k\right)\right)$$

s.t.,

$$\mathcal{C}_i\left(k, s_i^k\right) = \eta_i^k \times \left(\exp\left(\frac{s_i^k}{\mathcal{A}_i}\right) - 1\right) \text{ and } \sum_{k=1}^{m} s_i^k \le \mathcal{A}_i \tag{10}$$

where $\mathcal{C}_i\left(k, s_i^k\right)$ and η_i^k are the player i's cost function with the strategy s_i^k, and a cost control parameter for the task k, respectively. Traditionally, the exponential function is widely used in literature to model the consuming cost with assigned resource (Kim, 2014).

The goal of each players in N is to maximize all their profits; they select their strategies from a selfish motive. If some tasks needs high costs for sensing works, players in N do not contribute their resources for these non-profitable tasks, and these tasks can not be completed. At this time, all players pay a penalty cost ($\mathfrak{F}(\cdot)$), which is a great damage for players in N. Therefore, each individual player prefers that all tasks are completed, but also, they prefer that other payers contribute their resources to non-profitable tasks. Due to this reason, each player faces the decision of either making a small sacrifice from which all will benefit, or freeriding.

Usually, social dilemmas are situations in which the optimal decision of an individual contrasts with the optimal decision for the group. In game theory, this usually means games in which a dominant strategy leads to a Pareto inefficient equilibrium; the prisoner's dilemma is probably the most famous example (Archetti, 2009). In 1985, M. Diekmann first proposed the volunteer's dilemma game in the social sciences. In this game, each individual player prefers to avoid the cost of volunteering and exploit the benefit of the public good, but someone must volunteer and pay the cost of producing the good; if nobody volunteers, the cost paid is greater than the cost of volunteering (Diekmann, 1985).

In the traditional volunteer's dilemma game, a public good is produced if and only if at least one player volunteers to pay a cost. The basic model of N-person volunteer's dilemma is the following: each of N individuals can choose to volunteer (Volunteer) or not (Ignore). A public good is produced if and only if at least one individual volunteers (Archetti, 2009). Volunteering has a cost $c > 0$. Therefore, the individuals that volunteer have a payoff $1 - c$ and the ones that do not have a payoff 1. If nobody volunteers, the public good is not produced; everybody pays a cost $a > c$ (i.e., payoff $1 - a$). The fitness of the pure strategy Volunteer (W_V) is $W_V = 1 - c$ and the fitness of the pure strategy Ignore (W_I) is

$$W_I = \left(\gamma^{N-1} \times \left(1 - a\right)\right) + \left(1 - \gamma^{N-1}\right)$$

where γ is the probability of ignoring (not volunteering). The fitness of the mixed strategy is

$$W_{mix} = \left(\gamma \times W_I\right) + \left(\left(1 - \gamma\right) \times W_V\right).$$

The mixed-strategy equilibrium (γ_{eq}) can be found by equating the fitness of the two pure strategies; $\gamma_{eq} = \left(c / a\right)^{1/(N-1)}$. The volunteer's dilemma can be applied to many cases in the social sciences (Archetti, 2009).

For effective MCS operations, the IMCS scheme applies the concept of volunteer's dilemma to the OCF game model. When the amount of contribution for a specific task is less than its corresponding threshold, players make two decisions;

1. Whether to contribute their resources or not,
2. If they contribute, how much resource would be contributed.

Under dynamically changing MCS environments, these decisions should be made to effectively adapt to the current MCS condition. In order to adaptively implement this decision process, players can learn how to perform well by interacting with other players and dynamically adjust their decisions.

To decide adaptively above two questions, the IMCS scheme presents a novel dynamic learning approach. While time is ticking away, players obtain their payoffs ($\boldsymbol{U}_i\left(\cdot\right)$) as a consequence of their decisions. Based on this information, each player individually decides their strategy selection probability. If the task k has failed to be completed, the central server announces this situation. At this time, the player i in $\boldsymbol{N}$ selects his strategy (s_i^k); s_i^k can be defined as multiple amount levels of contributing resource, i.e., $s_i^k = \left\{s_i^{k(0)}, \ldots s_i^{k(\pm)}\right\}$ where $s_i^{k(0)}$ $(or\, s_i^{k(\pm)})$ means that the player i does not contribute his resource (*or* contributes $\pm$ more units of his resource). For the ($t+1$)th game round time, the selection probability for the $s_i^{k(d),1\le d\le\pm}$ strategy $\left(p^{t+1}\left(s_i^{k(d)}\right)\right)$ is estimated based on the propensity of $s_i^{k(d)}$ $\left(\mathcal{G}_i^{t+1}\left(s_i^{k(d)}\right)\right)$; it is described in the following;

$$p^{t+1}\left(s_i^{k(d)}\right) = e^{\frac{\mathcal{G}_i^{t+1}\left(s_i^{k(d)}\right)}{\psi}} \Bigg/ \sum_{h=1}^{\pm}\left(e^{\frac{\mathcal{G}_i^{t+1}\left(s_i^{k(h)}\right)}{\psi}}\right)$$

s.t.,

$$\begin{cases} \mathcal{G}_i^{t+1}\left(s_i^{k(d)}\right) = \left\{\left(\left(1-\xi\right)\times\mathcal{G}_i^t\left(s_i^{k(d)}\right)\right) + \left(\dfrac{\delta^k}{\pm}\times\left[\dfrac{\boldsymbol{U}_i^t\left(\boldsymbol{S}_i\right)-\boldsymbol{U}_i^{t-1}\left(\boldsymbol{S}_i\right)}{\boldsymbol{U}_i^{t-1}\left(\boldsymbol{S}_i\right)}\right]\right)\right\} & \textit{if } s_i^{k(d)} \in S_i \textit{ is selected at the time and } \delta^k = \sum_{l=1}^{n}\left(\dfrac{\zeta^k}{\zeta^l}\right) \\ \mathcal{G}_i^{t+1}\left(s_i^{k(f)}\right) = \left\{\left(1-\xi\right)\times\mathcal{G}_i^t\left(s_i^{k(f)}\right)\right\}, & \textit{otherwise, s.t.,} s_i^{k(f)} \neq s_i^{k(d)} \end{cases} \tag{11}$$

where ψ is a positive Boltzmann cooling parameter and ξ is the forgetting factor; it is essentially required when players face a game when the propensity adaptively jump changes over time. δ^k is the learning rate for the task k toward maximizing the utility function.

Considering the equations (9)-(11), the IMCS scheme can set the maximization problem. From the viewpoint of central server, the main interest is to maximize its total MCS revenue according to the dynamically adjusting $\mathcal{P}^k$ for each task. From the viewpoint of self-interesting individual smartphones, the major goal is to maximize their own payoff by selecting their strategies s_i^k. That is formally formulated as

$$\max_{\mathcal{P}^k}\left\{\boldsymbol{U}_0\left(\boldsymbol{S}_0\right) \mid \mathcal{P}^k \in \boldsymbol{S}_0, 1 \le k \le m\right\} \text{and}$$

$$\max_{s_i^k}\left\{\left[\boldsymbol{U}_i\left(\boldsymbol{S}_i\right) - \mathcal{F}\right] \mid s_i^k \in \boldsymbol{S}_i, 1 \le i \le n\right\}$$

s.t.,

$$\mathcal{F} = \sum_{k=1}^{m}\mathfrak{F}\left(k\right) \text{ and } \mathcal{C}_i\left(k, \mathcal{A}_i\right) \le \mathfrak{F}\left(k\right) \tag{12}$$

where $\mathfrak{F}\left(k\right)$ is the damage cost for everybody smartphone users if the task k has not been completed. Since the utilities of players are obtained from multiple tasks, players dynamically select different strategies for different tasks. For each task, smartphones invest their resources, and a central server assigns different *unit_prices*, individually. Therefore, players' interactive actions among various tasks is important in maximizing the players' own income, and improving the sensing performance of the MCS platform.

In the OCF game model, game players are selfish but cooperate with each other to effectively form coalitions to accomplish the tasks. At each round of OCF game operations, players periodically observe the other players' behaviors and coordinate explicitly by adjusting their strategies. In the scenario, coalitions are tasks that consists of multiple players in $\mathbf{N}$. They are willing to participate in tasks while automatically forming coalitions; they can be overlapping, and their contributions are rewarded by the central server. According to the *individual rationality*, they do not invest their resources to a specific task unless their work produces any profits. From the point of view of central server, a cost-loss can be caused if the participation level of players in $\mathbf{N}$ is less than some tasks' thresholds. Or a profit-loss can be caused when the participation level of players in $\mathbf{N}$ is much higher than some tasks' thresholds. Therefore, to approximate an optimized solution, the central server adaptively adjusts the sensing prices through any possible set of central server's price strategies to encourage players in participating or withdrawing some task works. During the step-by-step iteration, players individually adjust their strategies by using the dynamics of feedback-based repeated process, and attempt to guarantee the *group rationality*. Therefore, under widely diverse MCS situations, the main advantage of the IMCS scheme is a real-world practicality.

Summary

Nowadays, the exponential growth of smartphones creates a compelling paradigm of MCS. The MCS paradigm provides us various exciting and profitable applications. However, few studies addressed this issue, which significantly affects the functionality of MCS systems. The IMCS scheme is proposed as a novel MCS control algorithm based on the repeated OCF game model. The IMCS scheme is specifically for the real-world MCS system, where the smartphone users are paid off based on their contributions. Moreover, the IMCS scheme has incorporated the concept of volunteer's dilemma to maximize the total system performance.

CLOUD BASED SOCIAL NETWORK SERVICE (CSNS) SCHEME

According to the optimal mechanism design, the Cloud based Social Network Service (CSNS) scheme is developed as a novel cloud based Social Network Services (SNS) control scheme. To maximize the total system performance, social services have been adaptively executed in the cloud system while taking into account the social welfare. The CSNS scheme can capture the properties of SNS and Cloud Computing (CC), and provides an effective solution for the social cloud system.

Development Motivation

With the advance of the IoT, SNS has attracted billions of Internet users from all over the world in the past few years. SNS connects people to provide on-line communication and collaboration environment beyond the geographic limitations. It is considered to be a representative of the new generation Internet applications, and many specialized SNSs have emerged (Du, 2012). Usually, the main goal of SNS is to seek reciprocal value creation to increase the productivity, quality, and opportunities of online services. To satisfy this goal, SNS users will expect more application services to fulfill their needs beyond fundamental service functions. Therefore, how to link the needs of users and shape the designs for better service utilization is an important issue in SNS research fields (Hwang, 2012).

At present, Cloud Computing (CC) has been developed rapidly and becomes very common. Usually, *cloud* is used in science to describe a large agglomeration of objects that visually appear from a distance. In the Information and Communications Technology (ICT) field, it is a kind of Internet-based computing paradigm that provides shared processing resources and data to computers and other devices on demand. In particular, this paradigm represents a distributing computing model for enabling ubiquitous, convenient, on-demand network access to a shared pool of configurable computing resources. Therefore, CC technology provides flexible and scalable services without having the computing resources installed directly on SNS users' systems (Pan, 2014).

For the interoperability of the SNS and CC services, a new concept, Social Cloud (SC) was introduced based on the notion of resource and service collaboration. Despite the rapid development of the SC framework, there are some existing problems. One of most famous problems is to maximize the social welfare. It can be likened to the influence maximization problem (Chen, 2009). Consider the following scenario as an example. A CC system operator wants to provide cloud services for users of SNs. However, the CC system has a limited resource such that it can only select a small number of users to provide CC services. The CC provider wishes that these selected users would have strong relationship with their friends on the SN and share the profit of provided CC services. Therefore, through the ripple effect, a large population in the SN would be satisfied while maximizing their payoff. It's a good example of influence maximization. Simply, the influence maximization problem is the problem of detecting a set of influential users, who strongly influence the largest number of people in a SN (Chen, 2009).

Under widely dynamic SC system conditions, finding the best solution of the influence maximization problem is very challenging; it is a NP-hard problem (Kempe, 2003). The CSNS scheme focuses on the game theory to obtain an efficient solution for the influence maximization problem. Game theory is a decision-making process between independent decision-making players as they attempt to reach a joint decision that is acceptable to all participants. Facing to the modeling of decision-making issues, game theory has been regarded as an effective tool. After introducing the main concept of game theory, a lot of great outcomes have been achieved in a wider range of real life situations, such as political science, sociology, psychology, biology, and so on, where conflict and cooperation exist. Since the early 2010s, SNS and CC management issues have been added to this list (Kim, 2014)

In 1981, R. Myerson introduced the fundamental notion of optimal auction mechanism (Myerson, 1981). It is an applied branch of game theory which deals with how people act in auction and researches the properties of auction. Usually, auction design is represented by a mathematical game model; the game players are the buyers and sellers, and the action set of each player is a set of bid functions or reservation prices. The payoff of each player is the expected utility of that player under a combination of strategies (Kim, 2014). Optimal auction mechanism is a simple generalization of the second-price auction. Furthermore, it is dominant strategy incentive compatible and prior-free, i.e., it is not dependent on distributional assumptions (Myerson, 1981).

The Influential Maximization Problem for Social Clod Systems

Network diffusion formulates a scenario in which local interaction along edges in a graph can generate global cascades in network state. Such diffusion processes have attracted a significant amount of recent attention. Based on the network diffusion model, a conventional influence maximization problem aims to select some nodes so that the expected number of nodes influenced by selected nodes will be maximized (Kempe, 2003; Kim, 2013; Mihara, 2015). The CSNS scheme considers a SN as a directed graph $\mathbb{G} = \{\boldsymbol{V}, \boldsymbol{E}\}$ where $\boldsymbol{V}$ and $\boldsymbol{E}$ are the set of nodes and edges, respectively. Nodes represent the individuals in the SN and edges model the relationship between individuals. Mathematically, the influence maximization problem addresses the top $\mathcal{K}$ node set ($\boldsymbol{\mathcal{S}}$) from a graph $\mathbb{G}$ that satisfies

$$\boldsymbol{\mathcal{S}} = \arg\max_{\boldsymbol{\mathcal{S}} \subseteq V} \left\{ \mathfrak{F}\left(\boldsymbol{\mathcal{S}}, \mathbb{G}\right) \right\}, s.t., \left|\boldsymbol{\mathcal{S}}\right| = \mathcal{K} \tag{13}$$

where $\mathfrak{F}\left(\boldsymbol{\mathcal{S}}, \mathbb{G}\right) : \boldsymbol{\mathcal{S}} \times \boldsymbol{E} \rightarrow \mathbb{R}$ is a function of a node set $\boldsymbol{\mathcal{S}}$ that provides the expected number of the influenced nodes in the graph $\mathbb{G}$. Therefore, the influence maximization problem becomes an instance of a combinatorial optimization problem. According to the underlying graph structure, $\boldsymbol{\mathcal{S}}$ can be varied dynamically (Kim, 2013).

Nowadays, successful social media platforms are attractive not only for communication but also for information dissemination. This information diffusion in SN is regarded as an important mechanism that can improve social welfare. For example, an individual SN user needs to execute a computation complex application. Due to the limitation of embedded resource, some computation tasks can be offloaded to the CC system. When this user receives an outcome of CC service, it can be shared with his social friends. Therefore, from the viewpoint of CC provider, detecting influential users is important issue for effective and efficient SC operations. However, the CC provider cannot perfectly know the SN's topological structure. Therefore, the influence maximization problem is a very difficult problem (Chen, 2009; Deng, 2015; Jiao, 2012; Mihara, 2015).

In the CSNS scheme, a novel SC resource allocation scheme is developed to maximize the social welfare in SN. To avoid incurring the overhead of implementation complexity, the CSNS scheme leverages the optimal auction mechanism to allocate the cloud resource. To characterize the CSNS scheme, there is eight-tuple ($\mathbb{G}$, $\mathcal{CC}_p$, $f_{i \in V}$, $\mathbb{R}_i$, $\mathbb{A}$, $\mathfrak{W}$, $\boldsymbol{U}_{i \in V}$, $\mathbb{U}^{\mathcal{CC}_p}$).

1. $\mathbb{G}$ is a SN as a directed graph $\mathbb{G} = \{\boldsymbol{V}, \boldsymbol{E}\}$. $\boldsymbol{V} = \{1 \ldots n\}$ is a set of SN users, and $\boldsymbol{E} = \left\{e_1, \ldots e_m\right\}$ is a set of user's social relationships.
2. $\mathcal{CC}_p$ is a CC provider in the SC system.
3. $f_{i \in V}$ is the probability density function for the influence power of individual user i; the influence power can be measured as number of friends, who are connected in social relationships, and the weight of relationship.
4. $\mathbb{R}_i$ is the set of user i's friends; If the user j is a member of $\mathbb{R}_i$, the user i and the user j are connected in the SN. Based on their intimacy,

ψ_{ij} is denoted as the closeness of them; $\psi_{ij} \rightarrow [0,1]$ represents the weight of relationship. In this model, the profit sharing of CC service can be measured by ψ value. If there is no SN connection between the user i and j, $\psi_{ij} = 0$.

5. $\mathbb{A} = \left\{ \mathcal{A}_1 \ldots \mathcal{A}_j \ldots \mathcal{A}_{\mathcal{K}} \right\}$ is a finite resource allocation set for the selected $\mathcal{K}$ users where $j \in \boldsymbol{\mathcal{S}}$; $\mathcal{A}_j$ means the amount of allocated CC resource for the selected user j.
6. $\mathfrak{W}$ is the CC_p 's total computation resource amount for the CC services.
7. $\boldsymbol{U}_{i \in V}$ is the utility function to represent the payoff of the user i where $\boldsymbol{U}_i \rightarrow \mathbb{R}_+$. The social welfare of the SC system is defined as $\sum_{i \in V} \boldsymbol{U}_i$.
8. $\mathbb{U}^{CC_p}$ is the utility function of CC_p. It represents the social welfare of SC system.

Optimal Auction and Interactive Cost Control Mechanisms

When requesting the CC service, SN users can announce individually their social influence powers. Usually, the CC_p does not know the SN structure. Therefore, the CC_p 's problem is to infer what are the true influence powers of individual users from the announced information. The CSNS scheme shall assume that the announced influence power of each user can be described by a continuous probability distribution over a finite interval. Let ℓ_i $(or\, h_i)$ be the user i 's lowest (*or* highest) possible influence power level, i.e., $-\infty < \ell_i < h_i < +\infty$. The probability density function (f_i) for the user i 's influence power is s a continuous function on $\left[\ell_i, h_i\right]$. $\mathcal{F}_i$ will denote the cumulative distribution function corresponding to the density function f_i (Myerson, 1981);

$$\mathcal{F}_i\left(\varepsilon_i\right) = \int_{\ell_i}^{\varepsilon_i} f_i\left(\delta_i\right) d\delta_i, s.t., f_i\left(\delta_i\right) > 0 \text{ and } \mathcal{F}_i\left(\varepsilon_i\right) \rightarrow \left[0,1\right] \quad (14)$$

where $\varepsilon_i \in [\ell_i, h_i]$ is the user i's announce about his influence power. Therefore, $\mathcal{F}_i(\varepsilon_i)$ is the $\mathcal{CC}_p$'s assessment of the probability that user i has an influence power ε_i or less (Myerson, 1981). And then, denote

$$\mathbb{T} = [\ell_1, h_1] \times \ldots \times [\ell_i, h_i] \times \ldots \times [\ell_n, h_n]$$

be the set of all possible combinations of users' influence power announces (Myerson, 1981). The announced influence powers are stochastically independent random variables. Therefore, the joint density function on $\mathbb{T}$ for the announced influence power vector $\bullet = (\varepsilon_1, \cdots, \varepsilon_n)$ is defined as (Myerson, 1981);

$$\phi(\bullet) = \prod_{i \in V} f_i(\varepsilon_i) \tag{15}$$

During the SC system operation, individual users can define their utility function corresponding to the received CC service benefit minus the incurred cost to share the CC resource. Usually, CC services can be estimated as a concave benefit function, which provides monotone increasing value in proportion to the assigned CC resource amount. Therefore, the utility function ($\boldsymbol{U}_i$) of the user i is given by;

$$\boldsymbol{U}_i = \left\{\mathfrak{A}_i \times \mathcal{G}(i)\right\} + \left\{\left(\beta \times \mathfrak{A}_j \times \mathcal{G}(j)\right)^{1/(1+\psi_{ij})}\right\}$$

s.t.,

$$\mathfrak{A}_i = \left[\left(\cos\left(\frac{3\pi}{2} \times \frac{\mathcal{A}_i}{(\sigma \times \mathcal{M}_i)}\right)\right)^{1-\varepsilon_i} - \left(\zeta \times \left(\vartheta(\mathcal{A}_i) / \vartheta(\mathcal{M}_i)\right)\right)\right]$$

and

$$\mathcal{G}(i) = \begin{cases} 1, & if\ \mathcal{A}_i > 0 \\ 0, & otherwise \end{cases}, j \in \mathbb{R}_i \tag{16}$$

where $\mathcal{M}_i$ is the requested resource amount, and $\mathcal{A}_i$ is the assigned resource amount from the CC_p. σ is a modification factor, and ζ is a cost control parameter. $\vartheta\left(\mathcal{M}_i\right)$, $\vartheta\left(\mathcal{A}_i\right)$ are the energy consumption to execute $\mathcal{M}_i$ and $\mathcal{A}_i$ amount resources, respectively. In the CSNS scheme, $\vartheta\left(\cdot\right)$ is a linear function. β is a related payoff control parameter. In the equation (16), the first term means the payoff from the CC service, and the second term captures the profit sharing with social friends.

Based on the density functions f, and the utility function $\boldsymbol{U}$, the CC_p 's problem is to select some powerfully influential users in SNs to maximize the social welfare; it is a kind of influential maximization problem. In the CSNS scheme, the methodology of optimal auction mechanism is adopted to effectively solve this problem. The main feature of optimal auction mechanism is *direct revelation*. In a *direct revelation* mechanism, the SN users truthfully announce their influential powers to the CC_p; and the CC_p can determine who gets the CC service to maximize the social welfare. According to the optimal auction mechanism (Myerson, 1981), the expected utility ($\mathbb{U}^{CC_p}$) for the CC_p is defined as follows;

$$\mathbb{U}^{CC_p} = \int_{\mathbb{T}} \left(\mathfrak{A}_i + \sum_{\mathcal{A}_i \in \mathbb{A}} \left(\zeta \times \left(\vartheta\left(\mathcal{A}_i\right) \Big/ \vartheta\left(\mathcal{M}_i\right) \right) \right) \right) \times \phi\left(\bullet\right) d\bullet$$

s.t.,

$$\sum_{\mathcal{A}_i \in \mathbb{A}} \left(\mathcal{A}_i \times \mathcal{G}\left(i\right) \right) \leq \mathfrak{W} \tag{17}$$

To maximize the $\mathbb{U}^{CC_p}$, we should maximize the following mathematical formula by deciding the $\mathcal{G}\left(i\right)$ value.

$$\max_{\mathcal{G}(i \in V)} \left(\int_{\mathbb{T}} \left(\sum_{i \in V} \left(\varepsilon_i - \frac{1 - \mathcal{F}_i\left(\varepsilon_i\right)}{f_i\left(\varepsilon_i\right)} \right) \times \mathcal{G}\left(i\right) \right) \times \phi\left(\bullet\right) d\bullet \right)$$

s.t.,

$$\begin{cases} i \in \mathcal{S}, & if\ \mathcal{G}(i) = 1 \\ i \notin \mathcal{S}, & otherwise \end{cases} \text{ and } \sum_{\mathcal{A}_i \in \mathbb{A}} \left(\mathcal{A}_i \times \mathcal{G}(i) \right) \leq \mathfrak{W} \tag{18}$$

The cost of SC resource is chosen by the $\mathcal{CC}_p$, and the environmental response to this CC service cost serves as an input to the $\mathcal{CC}_p$. Therefore, the SN users and $\mathcal{CC}_p$ are connected in a feedback loop within the SC environment. To maximize the social welfare, the cost control parameter (ζ) should be adjusted adaptively. In order to implement the effective SC resource sharing algorithm, The CSNS scheme designs a time driven approach. Therefore, the CSNS scheme partitions the time-axis into equal intervals of length *unit_time*, and repeatedly applies the optimal auction mechanism in the step-by-step interactive online manner. At the end of each *unit_time* period, the CSNS scheme dynamically adjusts the value of ζ to make the SC system more responsive to the current environment. The adjusted ζ value for the $t+1$ time period (ζ^{t+1}) is given by;

$$\zeta^{t+1} = \max\left\{ \left[\left(\left(1-\alpha\right) \times \zeta^{t} \right) + \mathcal{H}\left(\mathcal{R}^{t}, \mathcal{R}^{t-1}, \theta, m \right) \right], 0 \right\}$$

s.t.,

$$\mathcal{H}\left(\mathcal{R}^{t}, \mathcal{R}^{t-1}, \theta, m \right) = \left(\theta \times \left(\left(\mathcal{R}^{t} - \mathcal{R}^{t-1} \right) \Big/ \left(\mathcal{R}^{t-1} + 1 \right) \right) \right)^{m} \tag{19}$$

where α is a weighted parameter and ζ^{t} is the ζ value at the time t period. θ, m are control parameters to adjust the CC service cost. The function $\mathcal{H}(\cdot)$ is defined as a cubic function. According to the equation (19), the $\mathcal{CC}_p$ adapts quite well to dynamic SN environments while improving the CC resource utilization.

The Main Steps of the CSNS Scheme

By adopting the optimal auction mechanism, the CSNS scheme can detect the most influential users in SNs. To improve the resource utilization, the cost of CC services is adaptively adjusted based on the dynamics of the interactive feedback process. At each time round, the user selection and CC cost adjustment are iteratively repeated. The CSNS scheme is described by the following seven major steps:

Step 1: Control parameters $\boldsymbol{V}$, $\boldsymbol{E}$, $\mathfrak{W}$, $\mathcal{M}$, α, β, σ, θ and m are given from the simulation scenario (refer to the Table I).

Step 2: At each time round (t), some users in SNs randomly request the CC services to execute their task applications.

Step 3: According to (14),(15) and (18), the CC_p dynamically select the top $\mathcal{K}$ users, i.e., $\mathcal{K} \ll n$, and allocate the CC resource to maximize the social welfare.

Step 4: Based on the current resource allocation information ($\mathbb{A}$), the cost control parameter (ζ) is adaptively adjusted using (19).

Step 5: In an entirely distributed manner, each individual users and the CC_p constantly calculate their payoffs using the equation (16) and (17), respectively.

Step 6: During the step-by-step iteration, previous decisions are adaptively adjusted based on the dynamics of repeated game process.

Step 7: Under the real-world SC environments, SN users and the CC_p are mutually dependent on each other to maximize the social welfare, and they constantly are self-monitoring the current SC system conditions; proceeds to Step 2 for the next iteration.

Summary

According to the basic idea of optimal auction mechanism, the CSNS scheme can effectively select the most influential users while maximizing the CC resource utilization. Based on the iterative feedback process, all control decisions are dynamically adjusted. Under diverse SC system environments, the CSNS scheme's approach is a more realistic methodology for finding an effective solution with practical assumptions.

REFERENCES

Alnwaimi, G., Vahid, S., & Moessner, K. (2015). Dynamic Heterogeneous Learning Games for Opportunistic Access in LTE-Based Macro/Femtocell Deployments. *IEEE Transactions on Wireless Communications*, *14*(4), 2294–2308. doi:10.1109/TWC.2014.2384510

An, J., Gui, X., Yang, J., Sun, Y., & He, X. (2015). Mobile Crowd Sensing for Internet of Things: A Credible Crowdsourcing Model in Mobile-Sense Service. *IEEE BigMM*, *2015*, 92–99.

Archetti, M. (2009). Cooperation as a volunteers dilemma and the strategy of conflict in public goods games. *Journal of Evolutionary Biology*, *22*(11), 2192–2200. doi:10.1111/j.1420-9101.2009.01835.x PMID:19732256

Chalkiadakis, G., Elkind, E., Markakis, E., Polukarov, M., & Jennings, N. R. (2010). Cooperative games with overlapping coalitions. *Journal of Artificial Intelligence Research*, *39*, 179–216.

Chaves, F. S., Abbas-Turki, M., Abou-Kandil, H., & Romano, J. M. T. (2013). Transmission power control for opportunistic QoS provision in wireless networks. *IEEE Transactions on Control Systems Technology*, *21*(2), 315–331. doi:10.1109/TCST.2011.2181080

Chen, W., Wang, Y., & Yang, S. (2009). Efficient influence maximization in social networks. *ACM KDD*, *2009*, 199–208.

Deng, X., Pan, Y., Wu, Y., & Gui, J. (2015). Credit Distribution and influence maximization in online social networks using node features. *IEEE FSKD*, *2015*, 2093–2100.

Di, B., Wang, T., Song, L., & Han, Z. (2013). Incentive mechanism for collaborative smartphone sensing using overlapping coalition formation games. *IEEE GLOBECOM*, *2013*, 1705–1710.

Diekmann, A. (1985). Volunteers dilemma. *The Journal of Conflict Resolution*, *29*(4), 605–610. doi:10.1177/0022002785029004003

Du, Z., Wang, Q., Fu, X., & Liu, Q. (2012). Integrated and flexible data management for cloud social network service platform on campus. *IEEE ICCSNT'2012*, (pp. 1241-1244).

Hwang, Y, C., & Shiau, W, C. (2012). Exploring Imagery-driven Service Framework on Social Network Service, *IEEE/ACM ASONAM'2012*, (pp. 1117-1122).

Jiao, Y., Wang, Y., Yuan, L., & Li, L. (2012). Cloud and SNS supported collaboration in AEC industry. *IEEE CSCWD, 2012*, 842–849.

Kempe, D., Kleinberg, J., & Tardos, É. (2003). Maximizing the spread of influence through a social network. *ACM SIGKDD, 2003*, 137–146.

Kim, J., Kim, S., & Yu, H. (2013). Scalable and parallelizable processing of influence maximization for large-scale social networks. *IEEE ICDE, 2013*, 266–277.

Kim, S. (2014). *Game Theory Applications in Network Design*. Philadelphia: IGI Global. doi:10.4018/978-1-4666-6050-2

Kouzayha, N., Taher, N. C., & Yacine, G. D. (2013). Towards a better support of machine type communication in LTE-networks: Analysis of random access mechanisms. *IEEE ICABME, 2013*, 57–60.

Li, J., Zhu, Y., & Yu, J. (2014). Load balance vs utility maximization in mobile crowd sensing: A distributed approach. *IEEE GLOBECOM, 2014*, 265–270.

Mihara, S., Tsugawa, S., & Ohsaki, H. (2015). Influence maximization problem for unknown social networks. *IEEE/ACM ASONAM'2015*, (pp. 1539-1546).

Myerson, R. (1981). Optimal auction design. *Mathematics of Operations Research*, *6*(1), 58–73. doi:10.1287/moor.6.1.58

Niyato, D., Wang, P., & Kim, D. I. (2014). Performance modeling and analysis of heterogeneous machine type communications. *IEEE Transactions on Wireless Communications*, 2836–2849.

Pan, Y., & Hu, N. (2014). Research on dependability of cloud computing systems. *IEEE ICRMS, 2014*, 435–439.

Salem, O., Liu, Y., & Mehaoua, A. (2013). Anomaly Detection in Medical Wireless Sensor Networks. *Journal for Corrosion Science and Engineering*, *70*, 272–284.

Sarcia, S. A. (2013). Timed strategic games; A new game theory for managing strategic plans in the time dimension. *IEEE CogSIMA, 2013*, 187–194.

Sheng, X., Tang, J., Xiao, X., & Xue, G., (2014). Leveraging GPS-Less Sensing Scheduling for Green Mobile Crowd Sensing. *IEEE Internet of Things Journal,* 328-336.

Yang, C. G., Li, J. D., & Li, W. Y. (2009). Joint rate and power control based on game theory in cognitive radio networks.*Fourth International Conference on Communications and Networking*, (pp. 1-5).

Zhan, Y., Wu, J., Wang, C., & Xie, J. (2012). On the Complexity and Algorithms of Coalition Structure Generation in Overlapping Coalition Formation Games. *IEEE ICTAI*, *2012*, 868–873.

Zhang, G., Li, A., Yang, K., Zhao, L., & Cheng, D. (2016). Optimal Power Control for Delay-Constraint Machine Type Communications over Cellular Uplinks. *IEEE Communications Letters*, *20*(6), 1168-1171.

KEY TERMS AND DEFINITIONS

Bluetooth: A wireless technology standard for exchanging data over short distances from fixed and mobile devices, and building personal area networks. Invented by telecom vendor Ericsson in 1994, it was originally conceived as a wireless alternative to RS-232 data cables. It can connect several devices, overcoming problems of synchronization.

Coalitional Games: Coalitions can be studied with tools of games. The most common solution concepts for coalitional games are the Core and the notion of internal and external stability. The latter describes a situation where no coalition member wants to leave the coalition and no non-member wants to join the coalition.

Machine Type Communication: Machine Type Communications (MTC) is a key segment in future cellular mobile packet data networks. Initial 3GPP efforts have focused on the ability to differentiate MTC devices, allowing operators to selectively handle MTC devices in congestion/overload situations. 3GPP has added additional specifications to integrate M2M communications into the overall network.

Mobile Crowd Sensing: A technique where a large group of individuals having mobile devices capable of sensing and computing collectively share data and extract information to measure, map, analyze, estimate or infer any processes of common interest.

Signal-to-Interference-Plus-Noise Ratio: In information theory and telecommunication engineering, the signal-to-interference-plus-noise ratio (SINR) is a quantity used to give theoretical upper bounds on channel capacity in wireless communication systems such as networks. Analogous to the SNR used often in wired communications systems, the SINR is defined as the power of a certain signal of interest divided by the sum of the interference power from all the other interfering signals and the power of some background noise.

Volunteer's Dilemma Game: Models a situation in which each of players faces the decision of either making a small sacrifice from which all will benefit, or freeriding. In this game, bystanders decide independently on whether to sacrifice themselves for the benefit of the group. If no one volunteers, everyone loses. The social phenomena of the bystander effect and diffusion of responsibility heavily relate to the volunteer's dilemma.

Related Readings

To continue IGI Global's long-standing tradition of advancing innovation through emerging research, please find below a compiled list of recommended IGI Global book chapters and journal articles in the areas of game theory, cloud computing, and cross domain applications. These related readings will provide additional information and guidance to further enrich your knowledge and assist you with your own research.

Achahbar, O., & Abid, M. R. (2015). The Impact of Virtualization on High Performance Computing Clustering in the Cloud. *International Journal of Distributed Systems and Technologies*, *6*(4), 65–81. doi:10.4018/IJDST.2015100104

Addya, S. K., Sahoo, B., & Turuk, A. K. (2015). Virtual Machine Placement Strategy for Cloud Data Center. In N. Rao (Ed.), *Enterprise Management Strategies in the Era of Cloud Computing* (pp. 261–287). Hershey, PA: IGI Global. doi:10.4018/978-1-4666-8339-6.ch012

Afghah, F., & Razi, A. (2014). Game Theoretic Study of Cooperative Spectrum Leasing in Cognitive Radio Networks. *International Journal of Handheld Computing Research*, *5*(2), 61–74. doi:10.4018/ijhcr.2014040104

Ahmad, A., & Ahmad, S. (2014). Radio Resource Management in Cognitive Radio Sensor Networks. In M. Rehmani & Y. Faheem (Eds.), *Cognitive Radio Sensor Networks: Applications, Architectures, and Challenges* (pp. 27–47). Hershey, PA: IGI Global. doi:10.4018/978-1-4666-6212-4.ch002

Akherfi, K., Harroud, H., & Gerndt, M. (2016). A Mobile Cloud Middleware to Support Mobility and Cloud Interoperability. *International Journal of Adaptive, Resilient and Autonomic Systems*, *7*(1), 41–58. doi:10.4018/IJARAS.2016010103

Related Readings

Akyuz, G. A., & Rehan, M. (2016). A Generic, Multi-Period and Multi-Partner Cost Optimizing Model for Cloud-Based Supply Chain. *International Journal of Cloud Applications and Computing*, *6*(2), 55–63. doi:10.4018/IJCAC.2016040106

Al-Mutairi, M. S. (2016). Fuzzy Optimal Approaches to 2-P Cooperative Games. *International Journal of Applied Industrial Engineering*, *3*(2), 22–35. doi:10.4018/IJAIE.2016070102

Al-nsour, S., Alryalat, H., & Alhawari, S. (2014). Integration between Cloud Computing Benefits and Customer Relationship Management (CRM) Processes to Improve Organizations Performance. *International Journal of Cloud Applications and Computing*, *4*(1), 1–14. doi:10.4018/ijcac.2014010101

Al-Somali, S., & Baghabra, H. (2016). Investigating the Determinants of IT Professionals Intention to Use Cloud-Based Applications and Solutions: An Extension of the Technology Acceptance. *International Journal of Cloud Applications and Computing*, *6*(3), 45–62. doi:10.4018/IJCAC.2016070104

Alavi, S. M., & Zhou, C. (2016). Auction-Based Resource Management in Multi-Cell OFDMA Networks. In C. Yang & J. Li (Eds.), *Game Theory Framework Applied to Wireless Communication Networks* (pp. 273–295). Hershey, PA: IGI Global. doi:10.4018/978-1-4666-8642-7.ch011

Alcarria, R., Martín, D., Robles, T., & Sánchez-Picot, Á. (2016). Enabling Efficient Service Distribution using Process Model Transformations. *International Journal of Data Warehousing and Mining*, *12*(1), 1–19. doi:10.4018/IJDWM.2016010101

Aljawarneh, S. A., & Yassein, M. O. (2016). A Conceptual Security Framework for Cloud Computing Issues. *International Journal of Intelligent Information Technologies*, *12*(2), 12–24. doi:10.4018/IJIIT.2016040102

Alkadi, I. (2016). Assessing Security with Regard to Cloud Applications in STEM Education. In L. Chao (Ed.), *Handbook of Research on Cloud-Based STEM Education for Improved Learning Outcomes* (pp. 260–276). Hershey, PA: IGI Global. doi:10.4018/978-1-4666-9924-3.ch017

AlZain, M. A., Li, A. S., Soh, B., & Pardede, E. (2015). Multi-Cloud Data Management using Shamirs Secret Sharing and Quantum Byzantine Agreement Schemes. *International Journal of Cloud Applications and Computing*, *5*(3), 35–52. doi:10.4018/IJCAC.2015070103

Amone, W. (2016). Game Theory. In B. Christiansen & E. Lechman (Eds.), *Neuroeconomics and the Decision-Making Process* (pp. 262–286). Hershey, PA: IGI Global. doi:10.4018/978-1-4666-9989-2.ch014

Aniyikaiye, J., & Udoh, E. (2016). Web Services Gateway: Taking Advantage of the Cloud. *International Journal of Grid and High Performance Computing*, *8*(1), 85–92. doi:10.4018/IJGHPC.2016010108

Arinze, B., Sylla, C., & Amobi, O. (2016). Cloud Computing for Teaching and Learning: Design Strategies. In L. Chao (Ed.), *Handbook of Research on Cloud-Based STEM Education for Improved Learning Outcomes* (pp. 159–171). Hershey, PA: IGI Global. doi:10.4018/978-1-4666-9924-3.ch011

Asadi, M., Agah, A., & Zimmerman, C. (2014). Applying Game Theory in Securing Wireless Sensor Networks by Minimizing Battery Usage. In A. Amine, O. Mohamed, & B. Benatallah (Eds.), *Network Security Technologies: Design and Applications* (pp. 58–73). Hershey, PA: IGI Global. doi:10.4018/978-1-4666-4789-3.ch004

Bathory, D. S. (2015). Proof for Evolution and Coming Out of Prison with Relational Dynamics. *International Journal of Applied Behavioral Economics*, *4*(1), 58–69. doi:10.4018/ijabe.2015010104

Beloudane, A., & Belalem, G. (2015). Towards an Efficient Management of Mobile Cloud Computing Services based on Multi Agent Systems. *Journal of Information Technology Research*, *8*(3), 59–72. doi:10.4018/JITR.2015070104

Ben Ayed, H. K., & Hamed, A. (2014). Toward Proactive Mobile Tracking Management. *International Journal of Information Security and Privacy*, *8*(4), 26–43. doi:10.4018/IJISP.2014100102

Benatia, I., Laouar, M. R., Bendjenna, H., & Eom, S. B. (2016). Implementing a Cloud-Based Decision Support System in a Private Cloud: The Infrastructure and the Deployment Process. *International Journal of Decision Support System Technology*, *8*(1), 25–42. doi:10.4018/IJDSST.2016010102

Benmerzoug, D. (2015). Towards AiP as a Service: An Agent Based Approach for Outsourcing Business Processes to Cloud Computing Services. *International Journal of Information Systems in the Service Sector*, *7*(2), 1–17. doi:10.4018/ijisss.2015040101

Bibi, S., Katsaros, D., & Bozanis, P. (2015). Cloud Computing Economics. In V. Díaz, J. Lovelle, & B. García-Bustelo (Eds.), *Handbook of Research on Innovations in Systems and Software Engineering* (pp. 125–149). Hershey, PA: IGI Global. doi:10.4018/978-1-4666-6359-6.ch005

Boehmer, W. (2015). Do We Need Security Management Systems for Data Privacy? In M. Gupta (Ed.), *Handbook of Research on Emerging Developments in Data Privacy* (pp. 263–299). Hershey, PA: IGI Global. doi:10.4018/978-1-4666-7381-6.ch013

Bouamama, S., & Belalem, G. (2015). The New Economic Environment to Manage Resources in Cloud Computing. *Journal of Information Technology Research, 8*(2), 34–49. doi:10.4018/jitr.2015040103

Bousia, A., Kartsakli, E., Antonopoulos, A., Alonso, L., & Verikoukis, C. (2016). Game Theoretic Infrastructure Sharing in Wireless Cellular Networks. In C. Yang & J. Li (Eds.), *Game Theory Framework Applied to Wireless Communication Networks* (pp. 368–398). Hershey, PA: IGI Global. doi:10.4018/978-1-4666-8642-7.ch014

Cesur-Kiliçaslan, S., & Işik, T. (2015). A General View of Poverty in Turkey as an Issue for Social Work in the Light of Behavioral Finance and Game Theory. In Z. Copur (Ed.), *Handbook of Research on Behavioral Finance and Investment Strategies: Decision Making in the Financial Industry* (pp. 25–37). Hershey, PA: IGI Global. doi:10.4018/978-1-4666-7484-4.ch002

Chahal, R. K., & Singh, S. (2015). Trust Calculation Using Fuzzy Logic in Cloud Computing. In K. Munir, M. Al-Mutairi, & L. Mohammed (Eds.), *Handbook of Research on Security Considerations in Cloud Computing* (pp. 127–172). Hershey, PA: IGI Global. doi:10.4018/978-1-4666-8387-7.ch007

Chaka, C. (2015). Personal Mobile Cloud Computing Affordances for Higher Education: One Example in South Africa. In N. Rao (Ed.), *Enterprise Management Strategies in the Era of Cloud Computing* (pp. 79–103). Hershey, PA: IGI Global. doi:10.4018/978-1-4666-8339-6.ch004

Chang, J., & Johnston, M. (2015). Approaches to Cloud Computing in the Public Sector: Case Studies in UK Local Government. In S. Aljawarneh (Ed.), *Advanced Research on Cloud Computing Design and Applications* (pp. 51–72). Hershey, PA: IGI Global. doi:10.4018/978-1-4666-8676-2.ch005

Chen, W., Wan, Y., Peng, B., & Amos, C. I. (2015). Genome Sequencing in the Cloud. In V. Chang, R. Walters, & G. Wills (Eds.), *Delivery and Adoption of Cloud Computing Services in Contemporary Organizations* (pp. 318–339). Hershey, PA: IGI Global. doi:10.4018/978-1-4666-8210-8.ch013

Choudhary, P. K., Mital, M., Sharma, R., & Pani, A. K. (2015). Cloud Computing and IT Infrastructure Outsourcing: A Comparative Study. *International Journal of Organizational and Collective Intelligence, 5*(4), 20–34. doi:10.4018/IJOCI.2015100103

Clanché, P., Jančařík, A., & Novotná, J. (2015). Off-Line Communication in Mathematics Using Mobile Devices. In M. Meletiou-Mavrotheris, K. Mavrou, & E. Paparistodemou (Eds.), *Integrating Touch-Enabled and Mobile Devices into Contemporary Mathematics Education* (pp. 147–176). Hershey, PA: IGI Global. doi:10.4018/978-1-4666-8714-1.ch007

Costan, A. A., Iancu, B., Rasa, P. C., Radu, A., Peculea, A., & Dadarlat, V. T. (2017). Intercloud: Delivering Innovative Cloud Services. In I. Hosu & I. Iancu (Eds.), *Digital Entrepreneurship and Global Innovation* (pp. 59–78). Hershey, PA: IGI Global. doi:10.4018/978-1-5225-0953-0.ch004

Das, R. (2014). A Game Theoretic Approach to Corporate Lending by the Banks in India. In B. Christiansen & M. Basilgan (Eds.), *Economic Behavior, Game Theory, and Technology in Emerging Markets* (pp. 271–288). Hershey, PA: IGI Global. doi:10.4018/978-1-4666-4745-9.ch015

Dawson, M. (2017). Exploring Secure Computing for the Internet of Things, Internet of Everything, Web of Things, and Hyperconnectivity. In M. Dawson, M. Eltayeb, & M. Omar (Eds.), *Security Solutions for Hyperconnectivity and the Internet of Things* (pp. 1–12). Hershey, PA: IGI Global. doi:10.4018/978-1-5225-0741-3.ch001

Dermentzi, E., Tambouris, E., & Tarabanis, K. (2016). Cloud Computing in eGovernment: Proposing a Conceptual Stage Model. *International Journal of Electronic Government Research, 12*(1), 50–68. doi:10.4018/IJEGR.2016010103

Diviacco, P., & Leadbetter, A. (2017). Balancing Formalization and Representation in Cross-Domain Data Management for Sustainable Development. In P. Diviacco, A. Leadbetter, & H. Glaves (Eds.), *Oceanographic and Marine Cross-Domain Data Management for Sustainable Development* (pp. 23–46). Hershey, PA: IGI Global. doi:10.4018/978-1-5225-0700-0.ch002

Dreher, P., Scullin, W., & Vouk, M. (2015). Toward a Proof of Concept Implementation of a Cloud Infrastructure on the Blue Gene/Q. *International Journal of Grid and High Performance Computing, 7*(1), 32–41. doi:10.4018/ijghpc.2015010103

Elkabbany, G. F., & Rasslan, M. (2017). Security Issues in Distributed Computing System Models. In M. Dawson, M. Eltayeb, & M. Omar (Eds.), *Security Solutions for Hyperconnectivity and the Internet of Things* (pp. 211–259). Hershey, PA: IGI Global. doi:10.4018/978-1-5225-0741-3.ch009

Elkhodr, M., Shahrestani, S., & Cheung, H. (2017). Internet of Things Research Challenges. In M. Dawson, M. Eltayeb, & M. Omar (Eds.), *Security Solutions for Hyperconnectivity and the Internet of Things* (pp. 13–36). Hershey, PA: IGI Global. doi:10.4018/978-1-5225-0741-3.ch002

Faheem, M., Kechadi, T., & Le-Khac, N. A. (2015). The State of the Art Forensic Techniques in Mobile Cloud Environment: A Survey, Challenges and Current Trends. *International Journal of Digital Crime and Forensics, 7*(2), 1–19. doi:10.4018/ijdcf.2015040101

Foti, M., & Vavalis, M. (2015). Intelligent Bidding in Smart Electricity Markets. *International Journal of Monitoring and Surveillance Technologies Research, 3*(3), 68–90. doi:10.4018/IJMSTR.2015070104

Gangwar, H., & Date, H. (2015). Exploring Information Security Governance in Cloud Computing Organisation. *International Journal of Applied Management Sciences and Engineering, 2*(1), 44–61. doi:10.4018/ijamse.2015010104

Garita, M. (2014). The Rationality of Dumping: The Case of Guatemala. In B. Christiansen & M. Basilgan (Eds.), *Economic Behavior, Game Theory, and Technology in Emerging Markets* (pp. 359–367). Hershey, PA: IGI Global. doi:10.4018/978-1-4666-4745-9.ch019

Georgalos, K. (2014). Playing with Ambiguity: An Agent Based Model of Vague Beliefs in Games. In D. Adamatti, G. Dimuro, & H. Coelho (Eds.), *Interdisciplinary Applications of Agent-Based Social Simulation and Modeling* (pp. 125–142). Hershey, PA: IGI Global. doi:10.4018/978-1-4666-5954-4.ch008

Ghafoor, K. Z., Mohammed, M. A., Abu Bakar, K., Sadiq, A. S., & Lloret, J. (2014). Vehicular Cloud Computing: Trends and Challenges. In J. Rodrigues, K. Lin, & J. Lloret (Eds.), *Mobile Networks and Cloud Computing Convergence for Progressive Services and Applications* (pp. 262–274). Hershey, PA: IGI Global. doi:10.4018/978-1-4666-4781-7.ch014

Grandinetti, L., Pisacane, O., & Sheikhalishahi, M. (2014). Cloud in Enterprises and Manufacturing. In *Pervasive Cloud Computing Technologies: Future Outlooks and Interdisciplinary Perspectives* (pp. 150–164). Hershey, PA: IGI Global. doi:10.4018/978-1-4666-4683-4.ch008

Grandinetti, L., Pisacane, O., & Sheikhalishahi, M. (2014). Cloud Computing and Operations Research. In *Pervasive Cloud Computing Technologies: Future Outlooks and Interdisciplinary Perspectives* (pp. 192–224). Hershey, PA: IGI Global. doi:10.4018/978-1-4666-4683-4.ch010

Hallappanavar, V. L., & Birje, M. N. (2017). Trust Management in Cloud Computing. In M. Dawson, M. Eltayeb, & M. Omar (Eds.), *Security Solutions for Hyperconnectivity and the Internet of Things* (pp. 151–183). Hershey, PA: IGI Global. doi:10.4018/978-1-5225-0741-3.ch007

Hashemi, S., Monfaredi, K., & Hashemi, S. Y. (2015). Cloud Computing for Secure Services in E-Government Architecture. *Journal of Information Technology Research, 8*(1), 43–61. doi:10.4018/JITR.2015010104

Hassan, B. M., Fouad, K. M., & Hassan, M. F. (2015). Keystroke Dynamics Authentication in Cloud Computing: A Survey. *International Journal of Enterprise Information Systems*, *11*(4), 99–120. doi:10.4018/IJEIS.2015100105

He, B., Tran, T. T., & Xie, B. (2014). Authentication and Identity Management for Secure Cloud Businesses and Services. In S. Srinivasan (Ed.), *Security, Trust, and Regulatory Aspects of Cloud Computing in Business Environments* (pp. 180–201). Hershey, PA: IGI Global. doi:10.4018/978-1-4666-5788-5.ch011

He, W., & Wang, F. (2015). A Hybrid Cloud Model for Cloud Adoption by Multinational Enterprises. *Journal of Global Information Management*, *23*(1), 1–23. doi:10.4018/jgim.2015010101

Huang, K., Li, M., Zhong, Z., & Zhao, H. (2016). Applications of Game Theory for Physical Layer Security. In C. Yang & J. Li (Eds.), *Game Theory Framework Applied to Wireless Communication Networks* (pp. 297–332). Hershey, PA: IGI Global. doi:10.4018/978-1-4666-8642-7.ch012

Hulsey, N. (2016). Between Games and Simulation: Gamification and Convergence in Creative Computing. In A. Connor & S. Marks (Eds.), *Creative Technologies for Multidisciplinary Applications* (pp. 130–148). Hershey, PA: IGI Global. doi:10.4018/978-1-5225-0016-2.ch006

Isaias, P., Issa, T., Chang, V., & Issa, T. (2015). Outlining the Issues of Cloud Computing and Sustainability Opportunities and Risks in European Organizations: A SEM Study. *Journal of Electronic Commerce in Organizations*, *13*(4), 1–25. doi:10.4018/JECO.2015100101

Islam, S., Fenz, S., Weippl, E., & Kalloniatis, C. (2016). Migration Goals and Risk Management in Cloud Computing: A Review of State of the Art and Survey Results on Practitioners. *International Journal of Secure Software Engineering*, *7*(3), 44–73. doi:10.4018/IJSSE.2016070103

Jasim, O. K., Abbas, S., El-Horbaty, E. M., & Salem, A. M. (2014). Cryptographic Cloud Computing Environment as a More Trusted Communication Environment. *International Journal of Grid and High Performance Computing*, *6*(2), 38–51. doi:10.4018/ijghpc.2014040103

Jasmine, K. S., & Sudha, M. (2015). Business Transformation though Cloud Computing in Sustainable Business. In F. Soliman (Ed.), *Business Transformation and Sustainability through Cloud System Implementation* (pp. 44–57). Hershey, PA: IGI Global. doi:10.4018/978-1-4666-6445-6.ch004

Ji, W., Chen, B., Chen, Y., Kang, S., & Zhang, S. (2016). Game Theoretic Analysis for Cooperative Video Transmission over Heterogeneous Devices: Mobile Communication Networks and Wireless Local Area Networks as a Case Study. In C. Yang & J. Li (Eds.), *Game Theory Framework Applied to Wireless Communication Networks* (pp. 427–456). Hershey, PA: IGI Global. doi:10.4018/978-1-4666-8642-7.ch016

Jouini, M., & Rabai, L. B. (2014). A Security Risk Management Metric for Cloud Computing Systems. *International Journal of Organizational and Collective Intelligence*, *4*(3), 1–21. doi:10.4018/ijoci.2014070101

Jouini, M., & Rabai, L. B. (2016). A Security Framework for Secure Cloud Computing Environments. *International Journal of Cloud Applications and Computing*, *6*(3), 32–44. doi:10.4018/IJCAC.2016070103

Kandil, A., El-Tantawy, O. A., El-Sheikh, S. A., & El-latif, A. M. (2016). Operation and Some Types of Soft Sets and Soft Continuity of (Supra) Soft Topological Spaces. In S. John (Ed.), *Handbook of Research on Generalized and Hybrid Set Structures and Applications for Soft Computing* (pp. 127–171). Hershey, PA: IGI Global. doi:10.4018/978-1-4666-9798-0.ch008

Kang, Y., & Yang, K. C. (2016). Analyzing Multi-Modal Digital Discourses during MMORPG Gameplay through an Experiential Rhetorical Approach. In B. Baggio (Ed.), *Analyzing Digital Discourse and Human Behavior in Modern Virtual Environments* (pp. 220–243). Hershey, PA: IGI Global. doi:10.4018/978-1-4666-9899-4.ch012

Kasemsap, K. (2015). Adopting Cloud Computing in Global Supply Chain: A Literature Review. *International Journal of Social and Organizational Dynamics in IT*, *4*(2), 49–62. doi:10.4018/IJSODIT.2015070105

Kasemsap, K. (2015). The Role of Cloud Computing Adoption in Global Business. In V. Chang, R. Walters, & G. Wills (Eds.), *Delivery and Adoption of Cloud Computing Services in Contemporary Organizations* (pp. 26–55). Hershey, PA: IGI Global. doi:10.4018/978-1-4666-8210-8.ch002

Kasemsap, K. (2016). The Fundamentals of Neuroeconomics. In B. Christiansen & E. Lechman (Eds.), *Neuroeconomics and the Decision-Making Process* (pp. 1–32). Hershey, PA: IGI Global. doi:10.4018/978-1-4666-9989-2.ch001

Katzis, K. (2015). Mobile Cloud Resource Management. In G. Mastorakis, C. Mavromoustakis, & E. Pallis (Eds.), *Resource Management of Mobile Cloud Computing Networks and Environments* (pp. 69–96). Hershey, PA: IGI Global. doi:10.4018/978-1-4666-8225-2.ch004

Khan, N., & Al-Yasiri, A. (2016). Cloud Security Threats and Techniques to Strengthen Cloud Computing Adoption Framework. *International Journal of Information Technology and Web Engineering, 11*(3), 50–64. doi:10.4018/IJITWE.2016070104

Kim, S. (2014). Bandwidth Management Algorithms by Using Game Models. In *Game Theory Applications in Network Design* (pp. 311–351). Hershey, PA: IGI Global. doi:10.4018/978-1-4666-6050-2.ch012

Kim, S. (2014). Game Models in Various Applications. In *Game Theory Applications in Network Design* (pp. 44–128). Hershey, PA: IGI Global. doi:10.4018/978-1-4666-6050-2.ch003

Kim, S. (2014). Game Theory for Network Security. In *Game Theory Applications in Network Design* (pp. 158–171). Hershey, PA: IGI Global. doi:10.4018/978-1-4666-6050-2.ch006

Kuada, E. (2017). Security and Trust in Cloud Computing. In M. Dawson, M. Eltayeb, & M. Omar (Eds.), *Security Solutions for Hyperconnectivity and the Internet of Things* (pp. 184–210). Hershey, PA: IGI Global. doi:10.4018/978-1-5225-0741-3.ch008

Kumar, D., Sahoo, B., & Mandal, T. (2015). Heuristic Task Consolidation Techniques for Energy Efficient Cloud Computing. In N. Rao (Ed.), *Enterprise Management Strategies in the Era of Cloud Computing* (pp. 238–260). Hershey, PA: IGI Global. doi:10.4018/978-1-4666-8339-6.ch011

Kumar, S. A. (2014). Organizational Control Related to Cloud. In S. Srinivasan (Ed.), *Security, Trust, and Regulatory Aspects of Cloud Computing in Business Environments* (pp. 234–246). Hershey, PA: IGI Global. doi:10.4018/978-1-4666-5788-5.ch014

Lai, W., Chang, T., & Lee, T. (2016). Distributed Dynamic Resource Allocation for OFDMA-Based Cognitive Small Cell Networks Using a Regret-Matching Game Approach. In C. Yang & J. Li (Eds.), *Game Theory Framework Applied to Wireless Communication Networks* (pp. 230–253). Hershey, PA: IGI Global. doi:10.4018/978-1-4666-8642-7.ch009

Li, W. H., Zhu, K., & Fu, H. (2017). Exploring the Design Space of Bezel-Initiated Gestures for Mobile Interaction. *International Journal of Mobile Human Computer Interaction, 9*(1), 16–29. doi:10.4018/IJMHCI.2017010102

Likavec, S., Osborne, F., & Cena, F. (2015). Property-based Semantic Similarity and Relatedness for Improving Recommendation Accuracy and Diversity. *International Journal on Semantic Web and Information Systems, 11*(4), 1–40. doi:10.4018/IJSWIS.2015100101

Limam, S., & Belalem, G. (2014). A Migration Approach for Fault Tolerance in Cloud Computing. *International Journal of Grid and High Performance Computing*, *6*(2), 24–37. doi:10.4018/ijghpc.2014040102

Lin, W., Yang, C., Zhu, C., Wang, J. Z., & Peng, Z. (2014). Energy Efficiency Oriented Scheduling for Heterogeneous Cloud Systems. *International Journal of Grid and High Performance Computing*, *6*(4), 1–14. doi:10.4018/IJGHPC.2014100101

Liu, C., Huang, K., Lee, Y., & Lai, K. (2015). Efficient Resource Allocation Mechanism for Federated Clouds. *International Journal of Grid and High Performance Computing*, *7*(4), 74–87. doi:10.4018/IJGHPC.2015100106

Martins, R. A., Kumar, K., Mukherjee, A., Nabin, M. H., & Bhattacharya, S. (2014). Decision-Making in Economics: Critical Lessons from Neurobiology. In B. Christiansen & M. Basilgan (Eds.), *Economic Behavior, Game Theory, and Technology in Emerging Markets* (pp. 46–56). Hershey, PA: IGI Global. doi:10.4018/978-1-4666-4745-9.ch004

Mayer, I., Bekebrede, G., Warmelink, H., & Zhou, Q. (2014). A Brief Methodology for Researching and Evaluating Serious Games and Game-Based Learning. In T. Connolly, T. Hainey, E. Boyle, G. Baxter, & P. Moreno-Ger (Eds.), *Psychology, Pedagogy, and Assessment in Serious Games* (pp. 357–393). Hershey, PA: IGI Global. doi:10.4018/978-1-4666-4773-2.ch017

Mezghani, K., & Ayadi, F. (2016). Factors Explaining IS Managers Attitudes toward Cloud Computing Adoption. *International Journal of Technology and Human Interaction*, *12*(1), 1–20. doi:10.4018/IJTHI.2016010101

Mihaljević, M. J., & Imai, H. (2014). Security Issues of Cloud Computing and an Encryption Approach. In M. Despotović-Zrakić, V. Milutinović, & A. Belić (Eds.), *Handbook of Research on High Performance and Cloud Computing in Scientific Research and Education* (pp. 388–408). Hershey, PA: IGI Global. doi:10.4018/978-1-4666-5784-7.ch016

Militano, L., Iera, A., Scarcello, F., Molinaro, A., & Araniti, G. (2016). Game Theoretic Approaches for Wireless Cooperative Content-Sharing. In C. Yang & J. Li (Eds.), *Game Theory Framework Applied to Wireless Communication Networks* (pp. 399–426). Hershey, PA: IGI Global. doi:10.4018/978-1-4666-8642-7.ch015

Mohammed, F., & Ibrahim, O. B. (2015). Drivers of Cloud Computing Adoption for E-Government Services Implementation. *International Journal of Distributed Systems and Technologies*, *6*(1), 1–14. doi:10.4018/ijdst.2015010101

Moura, J. A., Marinheiro, R. N., & Silva, J. C. (2014). Game Theory for Collaboration in Future Networks. In R. Trestian & G. Muntean (Eds.), *Convergence of Broadband, Broadcast, and Cellular Network Technologies* (pp. 94–123). Hershey, PA: IGI Global. doi:10.4018/978-1-4666-5978-0.ch005

Nagar, N., & Suman, U. (2014). Two Factor Authentication using M-pin Server for Secure Cloud Computing Environment. *International Journal of Cloud Applications and Computing*, *4*(4), 42–54. doi:10.4018/ijcac.2014100104

Nezarat, A., & Dastghaibifard, G. (2016). A Game Theoretic Method for Resource Allocation in Scientific Cloud. *International Journal of Cloud Applications and Computing*, *6*(1), 15–41. doi:10.4018/IJCAC.2016010102

Ng, A., Watters, P., & Chen, S. (2014). A Technology and Process Analysis for Contemporary Identity Management Frameworks. In M. Khosrow-Pour (Ed.), *Inventive Approaches for Technology Integration and Information Resources Management* (pp. 1–52). Hershey, PA: IGI Global. doi:10.4018/978-1-4666-6256-8.ch001

Orike, S., & Brown, D. (2016). Big Data Management: An Investigation into Wireless and Cloud Computing. *International Journal of Interdisciplinary Telecommunications and Networking*, *8*(4), 34–50. doi:10.4018/IJITN.2016100104

Ouf, S., & Nasr, M. (2015). Cloud Computing: The Future of Big Data Management. *International Journal of Cloud Applications and Computing*, *5*(2), 53–61. doi:10.4018/IJCAC.2015040104

Outanoute, M., Baslam, M., & Bouikhalene, B. (2015). Genetic Algorithm Learning of Nash Equilibrium: Application on Price-QoS Competition in Telecommunications Market. *Journal of Electronic Commerce in Organizations*, *13*(3), 1–14. doi:10.4018/JECO.2015070101

Patra, P. K., Singh, H., Singh, R., Das, S., Dey, N., & Victoria, A. D. (2016). Replication and Resubmission Based Adaptive Decision for Fault Tolerance in Real Time Cloud Computing: A New Approach. *International Journal of Service Science, Management, Engineering, and Technology*, *7*(2), 46–60. doi:10.4018/IJSSMET.2016040104

Peng, M., Sun, Y., Sun, C., & Ahmed, M. (2016). Game Theory-Based Radio Resource Optimization in Heterogeneous Small Cell Networks (HSCNs). In C. Yang & J. Li (Eds.), *Game Theory Framework Applied to Wireless Communication Networks* (pp. 137–183). Hershey, PA: IGI Global. doi:10.4018/978-1-4666-8642-7.ch006

Pereira, J. P. (2014). Simulation of Competition in NGNs with a Game Theory Model. In R. Trestian & G. Muntean (Eds.), *Convergence of Broadband, Broadcast, and Cellular Network Technologies* (pp. 216–243). Hershey, PA: IGI Global. doi:10.4018/978-1-4666-5978-0.ch010

Phelps, M., & Jennex, M. E. (2015). Ownership of Collaborative Works in the Cloud.[IJKM]. *International Journal of Knowledge Management, 11*(4), 35–51. doi:10.4018/IJKM.2015100103

Ramakrishna, V., & Dey, K. (2017). Mobile Application and User Analytics. In S. Mukherjea (Ed.), *Mobile Application Development, Usability, and Security* (pp. 231–259). Hershey, PA: IGI Global. doi:10.4018/978-1-5225-0945-5.ch011

Rao, N. R. (2015). Cloud Computing: An Enabler in Managing Natural Resources in a Country. In N. Rao (Ed.), *Enterprise Management Strategies in the Era of Cloud Computing* (pp. 155–169). Hershey, PA: IGI Global. doi:10.4018/978-1-4666-8339-6.ch007

Ratten, V. (2015). An Entrepreneurial Approach to Cloud Computing Design and Application: Technological Innovation and Information System Usage. In S. Aljawarneh (Ed.), *Advanced Research on Cloud Computing Design and Applications* (pp. 1–14). Hershey, PA: IGI Global. doi:10.4018/978-1-4666-8676-2.ch001

Ratten, V. (2015). Cloud Computing Technology Innovation Advances: A Set of Research Propositions. *International Journal of Cloud Applications and Computing, 5*(1), 69–76. doi:10.4018/ijcac.2015010106

Rawat, A., & Gambhir, S. (2015). Biometric: Authentication and Service to Cloud. In G. Deka & S. Bakshi (Eds.), *Handbook of Research on Securing Cloud-Based Databases with Biometric Applications* (pp. 251–268). Hershey, PA: IGI Global. doi:10.4018/978-1-4666-6559-0.ch012

Rawat, D. B., & Shetty, S. (2016). Game Theoretic Cloud-Assisted Opportunistic Spectrum Access in Cognitive Radio Networks. *International Journal of Grid and High Performance Computing, 8*(2), 94–110. doi:10.4018/IJGHPC.2016040106

Rindos, A., Vouk, M., & Jararweh, Y. (2014). The Virtual Computing Lab (VCL): An Open Source Cloud Computing Solution Designed Specifically for Education and Research. *International Journal of Service Science, Management, Engineering, and Technology, 5*(2), 51–63. doi:10.4018/ijssmet.2014040104

Ritzhaupt, A. D., Poling, N., Frey, C., Kang, Y., & Johnson, M. (2016). A Phenomenological Study of Games, Simulations, and Virtual Environments Courses: What Are We Teaching and How? *International Journal of Gaming and Computer-Mediated Simulations*, *8*(3), 59–73. doi:10.4018/IJGCMS.2016070104

Romm-Livermore, C., Raisinghani, M. S., & Rippa, P. (2016). The Politics of E-Learning: A Game Theory Analysis. *International Journal of Online Pedagogy and Course Design*, *6*(2), 1–14. doi:10.4018/IJOPCD.2016040101

Rusko, R. (2016). The Role of the Mixed Strategies and Selective Inflexibility in the Repeated Games of Business: Multiple Case Study Analysis. In B. Christiansen & E. Lechman (Eds.), *Neuroeconomics and the Decision-Making Process* (pp. 132–146). Hershey, PA: IGI Global. doi:10.4018/978-1-4666-9989-2.ch008

Salama, M., Zeid, A., Shawish, A., & Jiang, X. (2014). A Novel QoS-Based Framework for Cloud Computing Service Provider Selection. *International Journal of Cloud Applications and Computing*, *4*(2), 48–72. doi:10.4018/ijcac.2014040104

Santos, J. L. (2015). An Agent-Based Model of Insurance and Protection Decisions on IT Systems. *International Journal of Agent Technologies and Systems*, *7*(3), 1–17. doi:10.4018/IJATS.2015070101

Seçilmiş, İ. E. (2014). A Survey of Game Theory Applications in Turkey. In B. Christiansen & M. Basilgan (Eds.), *Economic Behavior, Game Theory, and Technology in Emerging Markets* (pp. 155–168). Hershey, PA: IGI Global. doi:10.4018/978-1-4666-4745-9.ch009

Shen, Y., Li, Y., Wu, L., Liu, S., & Wen, Q. (2014). Cloud Computing Overview. In Y. Shen, Y. Li, L. Wu, S. Liu, & Q. Wen (Eds.), *Enabling the New Era of Cloud Computing: Data Security, Transfer, and Management* (pp. 1–24). Hershey, PA: IGI Global. doi:10.4018/978-1-4666-4801-2.ch001

Shi, Z., & Beard, C. (2014). QoS in the Mobile Cloud Computing Environment. In J. Rodrigues, K. Lin, & J. Lloret (Eds.), *Mobile Networks and Cloud Computing Convergence for Progressive Services and Applications* (pp. 200–217). Hershey, PA: IGI Global. doi:10.4018/978-1-4666-4781-7.ch011

Singh, A., Dutta, K., & Singh, A. (2014). Resource Allocation in Cloud Computing Environment using AHP Technique. *International Journal of Cloud Applications and Computing*, *4*(1), 33–44. doi:10.4018/ijcac.2014010103

Singh, J., & Kumar, V. (2014). Multi-Disciplinary Research Issues in Cloud Computing. *Journal of Information Technology Research*, *7*(3), 32–53. doi:10.4018/jitr.2014070103

Sivagurunathan, S., & Swasthimathi, L. S. (2016). Cloud Computing Applications in Education through E-Governance: An Indian Perspective. In Z. Mahmood (Ed.), *Cloud Computing Technologies for Connected Government* (pp. 247–268). Hershey, PA: IGI Global. doi:10.4018/978-1-4666-8629-8.ch010

Stennikov, V., Penkovskii, A., & Khamisov, O. (2016). Problems of Modeling and Optimization of Heat Supply Systems: Bi-Level Optimization of the Competitive Heat Energy Market. In P. Vasant & N. Voropai (Eds.), *Sustaining Power Resources through Energy Optimization and Engineering* (pp. 54–75). Hershey, PA: IGI Global. doi:10.4018/978-1-4666-9755-3.ch003

Stranacher, K., Tauber, A., Zefferer, T., & Zwattendorfer, B. (2014). The Austrian Identity Ecosystem: An E-Government Experience. In A. Ruiz-Martinez, R. Marin-Lopez, & F. Pereniguez-Garcia (Eds.), *Architectures and Protocols for Secure Information Technology Infrastructures* (pp. 288–309). Hershey, PA: IGI Global. doi:10.4018/978-1-4666-4514-1.ch011

Subramanian, T., & Savarimuthu, N. (2016). Cloud Service Evaluation and Selection Using Fuzzy Hybrid MCDM Approach in Marketplace. *International Journal of Fuzzy System Applications*, *5*(2), 118–153. doi:10.4018/IJFSA.2016040108

Suthakar, K. I., & Devi, M. K. (2016). Resource Scheduling for Big Data on Cloud: Scheduling Resources. In R. Kannan, R. Rasool, H. Jin, & S. Balasundaram (Eds.), *Managing and Processing Big Data in Cloud Computing* (pp. 185–205). Hershey, PA: IGI Global. doi:10.4018/978-1-4666-9767-6.ch013

Suwais, K. (2014). Assessing the Utilization of Automata in Representing Players Behaviors in Game Theory. *International Journal of Ambient Computing and Intelligence*, *6*(2), 1–14. doi:10.4018/IJACI.2014070101

Takabi, H., Zargar, S. T., & Joshi, J. B. (2014). Mobile Cloud Computing and Its Security and Privacy Challenges. In D. Rawat, B. Bista, & G. Yan (Eds.), *Security, Privacy, Trust, and Resource Management in Mobile and Wireless Communications* (pp. 384–407). Hershey, PA: IGI Global. doi:10.4018/978-1-4666-4691-9.ch016

Thomas, M. V., & Chandrasekaran, K. (2016). Identity and Access Management in the Cloud Computing Environments. In G. Kecskemeti, A. Kertesz, & Z. Nemeth (Eds.), *Developing Interoperable and Federated Cloud Architecture* (pp. 61–90). Hershey, PA: IGI Global. doi:10.4018/978-1-5225-0153-4.ch003

Tomaiuolo, M. (2014). Trust Management and Delegation for the Administration of Web Services. In I. Portela & F. Almeida (Eds.), *Organizational, Legal, and Technological Dimensions of Information System Administration* (pp. 18–37). Hershey, PA: IGI Global. doi:10.4018/978-1-4666-4526-4.ch002

Truong, D. (2015). Efficiency and Risk Management Models for Cloud-Based Solutions in Supply Chain Management. *International Journal of Business Analytics*, *2*(2), 14–30. doi:10.4018/IJBAN.2015040102

Tuncalp, D. (2015). Management of Privacy and Security in Cloud Computing: Contractual Controls in Service Agreements. In V. Chang, R. Walters, & G. Wills (Eds.), *Delivery and Adoption of Cloud Computing Services in Contemporary Organizations* (pp. 409–434). Hershey, PA: IGI Global. doi:10.4018/978-1-4666-8210-8.ch017

Udoh, E., Patterson, B., & Cordle, S. (2016). Using the Balanced Scorecard Approach to Appraise the Performance of Cloud Computing. *International Journal of Grid and High Performance Computing*, *8*(1), 50–57. doi:10.4018/IJGHPC.2016010104

Umar, R., & Mesbah, W. (2015). Throughput-Efficient Spectrum Access in Cognitive Radio Networks: A Coalitional Game Theoretic Approach. In N. Kaabouch & W. Hu (Eds.), *Handbook of Research on Software-Defined and Cognitive Radio Technologies for Dynamic Spectrum Management* (pp. 454–477). Hershey, PA: IGI Global. doi:10.4018/978-1-4666-6571-2.ch017

Voderhobli, K. (2015). The Need for Traffic Based Virtualisation Management for Sustainable Clouds. *International Journal of Organizational and Collective Intelligence*, *5*(4), 8–19. doi:10.4018/IJOCI.2015100102

Vujin, V., Simić, K., & Kovačević, B. (2014). Digital Identity Management in Cloud. In M. Despotović-Zrakić, V. Milutinović, & A. Belić (Eds.), *Handbook of Research on High Performance and Cloud Computing in Scientific Research and Education* (pp. 56–81). Hershey, PA: IGI Global. doi:10.4018/978-1-4666-5784-7.ch003

Wang, C., Wei, H., Bennis, M., & Vasilakos, A. V. (2016). Game-Theoretic Approaches in Heterogeneous Networks. In C. Yang & J. Li (Eds.), *Game Theory Framework Applied to Wireless Communication Networks* (pp. 88–102). Hershey, PA: IGI Global. doi:10.4018/978-1-4666-8642-7.ch004

Wenge, O., Schuller, D., Rensing, C., & Steinmetz, R. (2014). On Developing Fair and Orderly Cloud Markets: QoS- and Security-Aware Optimization of Cloud Collaboration. *International Journal of Organizational and Collective Intelligence*, *4*(3), 22–43. doi:10.4018/ijoci.2014070102

Windsor, D. (2014). Business Ethics in Emerging Economies: Identifying Game-Theoretic Insights for Key Issues. In B. Christiansen & M. Basilgan (Eds.), *Economic Behavior, Game Theory, and Technology in Emerging Markets* (pp. 30–45). Hershey, PA: IGI Global. doi:10.4018/978-1-4666-4745-9.ch003

Windsor, D. (2015). Game-Theoretic Insights Concerning Key Business Ethics Issues Occurring in Emerging Economies. In D. Palmer (Ed.), *Handbook of Research on Business Ethics and Corporate Responsibilities* (pp. 34–55). Hershey, PA: IGI Global. doi:10.4018/978-1-4666-7476-9.ch003

Wu, D., & Cai, Y. (2016). Coalition Formation Game for Wireless Communications. In C. Yang & J. Li (Eds.), *Game Theory Framework Applied to Wireless Communication Networks* (pp. 28–62). Hershey, PA: IGI Global. doi:10.4018/978-1-4666-8642-7.ch002

Wu, J., Ding, F., Xu, M., Mo, Z., & Jin, A. (2016). Investigating the Determinants of Decision-Making on Adoption of Public Cloud Computing in E-government. *Journal of Global Information Management*, *24*(3), 71–89. doi:10.4018/JGIM.2016070104

Xu, S., & Xia, C. (2016). Resource Allocation for Device-to-Device Communications in LTE-A Network: A Stackelberg Game Theory Approach. In C. Yang & J. Li (Eds.), *Game Theory Framework Applied to Wireless Communication Networks* (pp. 212–229). Hershey, PA: IGI Global. doi:10.4018/978-1-4666-8642-7.ch008

Xu, X., Gao, R., Li, M., & Wang, Y. (2016). Interference Mitigation with Power Control and Allocation in the Heterogeneous Small Cell Networks. In C. Yang & J. Li (Eds.), *Game Theory Framework Applied to Wireless Communication Networks* (pp. 103–136). Hershey, PA: IGI Global. doi:10.4018/978-1-4666-8642-7.ch005

Xu, Y., Wang, J., & Wu, Q. (2016). Distributed Learning of Equilibria with Incomplete, Dynamic, and Uncertain Information in Wireless Communication Networks. In C. Yang & J. Li (Eds.), *Game Theory Framework Applied to Wireless Communication Networks* (pp. 63–86). Hershey, PA: IGI Global. doi:10.4018/978-1-4666-8642-7.ch003

Yaacoub, E., Ghazzai, H., & Alouini, M. (2016). A Game Theoretic Framework for Green HetNets Using D2D Traffic Offload and Renewable Energy Powered Base Stations. In C. Yang & J. Li (Eds.), *Game Theory Framework Applied to Wireless Communication Networks* (pp. 333–367). Hershey, PA: IGI Global. doi:10.4018/978-1-4666-8642-7.ch013

Yahaya, M. O. (2017). On the Role of Game Theory in Modelling Incentives and Interactions in Mobile Distributed Systems. In K. Munir (Ed.), *Security Management in Mobile Cloud Computing* (pp. 92–120). Hershey, PA: IGI Global. doi:10.4018/978-1-5225-0602-7.ch005

Yao, Y. (2015). Cloud Computing: A Practical Overview Between Year 2009 and Year 2015. *International Journal of Organizational and Collective Intelligence*, *5*(3), 32–43. doi:10.4018/ijoci.2015070103

Yao, Y. (2016). Emerging Cloud Computing Services: A Brief Opinion Article. *International Journal of Organizational and Collective Intelligence*, *6*(4), 98–102. doi:10.4018/IJOCI.2016100105

Yarlikas, S., & Bilgen, S. (2014). Measures for Cloud Computing Effectiveness Assessment. *International Journal of Cloud Applications and Computing*, *4*(3), 20–43. doi:10.4018/ijcac.2014070102

Zardari, M. A., & Jung, L. T. (2016). Classification of File Data Based on Confidentiality in Cloud Computing using K-NN Classifier. *International Journal of Business Analytics*, *3*(2), 61–78. doi:10.4018/IJBAN.2016040104

Zhong, W., Wang, J., & Tao, M. (2016). Potential Games and Its Applications to Wireless Networks. In C. Yang & J. Li (Eds.), *Game Theory Framework Applied to Wireless Communication Networks* (pp. 1–27). Hershey, PA: IGI Global. doi:10.4018/978-1-4666-8642-7.ch001

Zhou, Y., Huang, L., Tian, L., & Shi, J. (2016). Game Theory-Based Coverage Optimization for Small Cell Networks. In C. Yang & J. Li (Eds.), *Game Theory Framework Applied to Wireless Communication Networks* (pp. 184–211). Hershey, PA: IGI Global. doi:10.4018/978-1-4666-8642-7.ch007

About the Author

Sungwook Kim received the BS, MS degrees in computer science from the Sogang University, Seoul, in 1993 and 1995, respectively. In 2003, he received the Ph.D degree in computer science from the Syracuse University, Syracuse, New York, supervised by Prof. Pramod K. Varshney in 2003. He has held faculty positions at the department of Computer Science of ChoongAng University, Seoul. In 2006, he returned to his Alma Mater, Sogang University, where he is currently a Professor of Department of Computer Science & Engineering, and is a research director of the internet communication control research laboratory. His current research interests are in game theory and network design applications.

Index

B

Baseband Processing 16, 75, 77
Bayesian inference 111-112, 144, 146
Behavioral Game Theory 103, 106, 141, 146
Bitcoin 78, 87-99
Bitcoin mining 78, 87, 89-91, 93, 97-99
Bluetooth 7, 180, 200

C

cloud computing 2, 8, 11, 13-14, 25-26, 38-39, 41, 50-51, 59, 71-72, 74-77, 142, 190, 199
CloudIoT 2, 11
Cloud Radio Access Network (C-RAN) 13-14, 60, 77
coalitional games 181, 200
Cognitive Hierarchy Theory 146
Cognitive Radio 10, 72, 74, 119-120, 142-144, 146, 149-151, 154, 168-169, 200
computation offloading 38-44, 49, 53, 72, 75, 77
CSMA 116, 146

F

Fog Computing 4, 9, 11, 59-61, 72
Future Internet 2, 11, 13, 39, 49

G

Gaussian Noise 154, 170
Green networking 170

H

heterogeneous network system 79, 81, 99

I

incentive compatibility 86, 99, 164
individual rationality 21, 86, 93, 99, 178, 189
Internet of Things 1, 5, 8-11, 38, 71-72, 75, 79, 109, 142-143, 145-146, 198-199

K

Kalai-Smorodinsky Bargaining Solution 20, 73, 77

M

M2M Communications 11, 200
Machine Type Communication 172, 199-200
Mobile Cloud IoT 13, 77
Mobile Crowd Sensing 6-8, 11, 171, 180, 184, 198-200

N

Nash Bargaining Solution 20, 64, 77, 89, 169, 178

P

Pareto Optimality 21, 93-94, 164, 170, 178
power control 8, 72-73, 102-103, 105, 108, 130, 134-135, 137, 142-145, 147, 161, 172-173, 175, 177, 198, 200

Principal-Agent Problem 83, 99
Proof-of-Work (PoW) 89, 99

Q

Quality of Service 3, 12, 14, 28, 38, 60, 172

R

Radio Frequency 5, 26, 148-149, 170
Revelation Principle 86, 99
RFID 120, 147
Rubinstein-Stahl model 18, 24, 46-47, 77

S

SHA-256 Hashing 90, 100
Shapley Value 30, 32, 73, 77
signal-to-interference-plus-noise ratio 173, 200
Social Cloud 6, 8-9, 13, 50-53, 72, 77, 190
Social Network Service 6, 9, 12, 50, 189, 198

T

TDMA 151, 170
Tragedy of the Commons 52, 77

U

uncertainty 26, 80, 109-112, 114-115, 119, 141, 147
Underwater Sensor Networks 5, 9-10, 12

V

Virtual Currencies 88, 100
volunteer's dilemma game 186, 201

W

Wi-Fi networks 80-83, 100

Printed in the United States
By Bookmasters